Praise for *Mud, Microbes, and Medicine*

"A fascinating memoir—brilliant, bold, and original . . . a fearless tale of reinvention and resilience. You're about to enjoy eight delightful books fused into one fascinating autobiography. You'll read about life on remote Pacific Islands, low-tech ways of raising children, learning from one's mistakes, how to start out naïve and still succeed, and how to morph from an impoverished powerless student into an affluent powerful executive."

—Jared Diamond, Pulitzer Prize–winning author of *Guns, Germs, and Steel*

"What a rip-roaring gem of a text *Mud, Microbes, and Medicine* is, all the way from Safari Barbie in the '70s to the latest trip back to the real world there in the Southwestern Pacific in 2019. A gripping read. Congratulations to Elizabeth Aden for all her very hard work in getting it all together (in life, and on paper!)."

—Kirk W. Huffman, PhD, anthropologist, research associate at Australian Museum, curator and honorary curator of the Vanuatu Cultural Centre, and author of *Arts of Vanuatu*

"Elizabeth Aden's journey from naïve young researcher in Vanuatu to the powerhouses of Silicon Valley is wise, fascinating, funny, brutally honest, and beautifully written. Nearly every page has an insight about the different cultures, South Pacific to biotech boardrooms, that hits like a revelation. The book is also a moving story about the personal efforts and sacrifices behind the development of drugs for diseases like transplant medicine, and the challenges that continue to face women in science."

—Sheri T. Joseph, *USA Today* best-selling author of *Edge of the Known World*, winner in the 2024 American Fiction Awards for Best New Fiction

"*Mud, Microbes, and Medicine* is an engaging read that grants you the confidence to believe in yourself and follow where your heart and mind lead you."

—Carolyn Feamster, biopharma executive

"Life is an adventure. And few would dispute that Aden's life as a cultural anthropologist, infectious disease epidemiologist, pharmaceutical executive, and author would not qualify as adventuresome. *Mud, Microbes, and Medicine* takes on an intensely personal autobiographical retelling of Dr. Aden's life in science and industry. The lessons she learned along the way—conveniently summarized at the end of each chapter—make fascinating reading."

—George W. Rutherford, MD, MA, professor of epidemiology, preventive medicine, pediatrics, and history at University of California, San Francisco

"*In Mud, Microbes, and Medicine*, Elizabeth Aden takes us on a deeply personal yet universally relatable journey of self-discovery, courage, and resilience. With honesty, humor, and wit, she reflects on the complexities of relationships, societal expectations, and the often messy process of finding one's own path. It's a tale that speaks to anyone who has ever felt trapped, undervalued, or unsure of their future, showing that with faith, determination, and a bit of humor, we can rewrite our stories and embrace the freedom to live life on our own terms."

—Kim Sorrelle, author of *Love Is* and TV and podcast host

"Aden takes us on one of the wildest rides imaginable, surpassing even the thrills of political thrillers and action-packed adventures. Her nonfiction account of cultural shock and self-discovery in Melanesia is both riveting and eye-opening. The lessons she learns from her missteps offer inspiring insights, making her journey a remarkable recipe for success and personal growth. A must-read for anyone seeking adventure and inspiration!"

—Nancy Orlen Weber, author of *The Life of a Psychic Detective*, *All Nature Speaks*, and *The Soul Detective*

"Aden's memoir of curiosity and inquiry spans the globe. Her experiences are described in a timeless playbook for everyone navigating the uncertainty and opportunity of life. She sums it up best in her own words: 'When there are no rules, you get to write them in your favor.'"

—Pat Wetzel, host of the Bump in the Road podcast and author of the *Bump in the Road* book series

"*Mud, Microbes, and Medicine* is a riveting read of the personal and professional evolution of a scientist during fifty-plus years of risk-taking and continued learning in a man's world. Her combination of epidemiology and anthropology yielded important contributions to the natural history of hepatitis B virus infection."

—Barbara G Werner, PhD, FIDSA, retired public health infectious disease laboratory scientist and associated professor of medicine at Tufts University School of Medicine and UMass Medical School

"What a mixture of intrigue and science and lessons about living! Aden's honesty is very refreshing and fed my curiosity about what next decision she would make and where it would take her. It shows the kind of creative, tedious, and risky work that some scientists do. This book is a fast and enjoyable read."

—Mary Lang Sollinger, author of *From Inspiration to Activism: A Personal Journey Through Obama's Presidential Campaign*

"Anyone who has ventured alone into an unfamiliar culture, or anyone who has longed for this kind of adventure, will find *Mud, Microbes, and Medicine* a page-turner. Ms. Aden's journey of self-discovery is a chronicle of feminist pathfinding over the last fifty years. Her journey from fieldwork among the hospitable ni-Tamambo on the island of Malo in the New Hebrides to the not-always-so-hospitable halls of corporate decision-making shows ingenuity and courage, creativity and genuine grit. Her significant accomplishments in medical research are inspiring to those willing to aim high, take the path less traveled, and meet obstacles with persistence, self-awareness, and humor."

—J. Suzanne Cox, cross-cultural market strategist and former executive VP of marketing at American National Red Cross

MUD, MICROBES & MEDICINE

MUD, MICROBES & MEDICINE

How a Curious Anthropologist Got to the Boardroom

Elizabeth Reed Aden, PhD

SWP

SHE WRITES PRESS

Published 2026 by
She Writes Press, an imprint of The Stable Book Group

32 Court Street, Suite 2109
Brooklyn, NY 11201
https://shewritespress.com
Library of Congress Control Number: 2026900605
ISBN: 979-8-89636-038-4
eISBN: 979-8-89636-039-1

Interior Designer: Tabitha Lahr

Printed in the United States

Names and identifying characteristics have been changed to protect the privacy of certain individuals.

To the many women who helped, educated, and mentored me and gave me opportunities:
Eunice Mays Boyd, Gladys James, Scott Mazzur, Cyndie Mertens, Valerie Mayer, Karen Campbell, Pam Kirby, and Marilyn Lucia

To the men who opened doors and gave me the space to succeed:
Jonathan Knowles, Robert Booth, and Mel Aden

To my advisers, who made this adventure possible:
Ward Goodenough
who had faith in my ability to succeed and the courage to let me try
and
Barry Blumberg
who provided the means and opportunity, and who had the resilience to see it through

To the ni-Tamambo,
who graciously accepted me, forgave me, taught me, and welcomed me over the years

Thank you, all

Contents

List of Black-and-White Figures

Appendices:

Color Plates

Foreword

Dr. Elizabeth Aden's memoir, *Mud, Microbes, and Medicine: How a Curious Anthropologist Got to the Boardroom*, is a down-to-earth, vivid, honest, and engaging story of a woman's experiences as an anthropologist who transitioned into the world of business.

Dr. Aden takes the reader through a variety of settings: a failed marriage, graduate school, ethnographic medical research on a remote island in the South Pacific, and then, after leaving academia, various positions in the business world including top-level jobs in the pharmaceutical industry, and finally being a mom in a successful second marriage. We meet a variety of characters along the way: a loving but difficult mother, a Nobel Prize–winning researcher, university professors who are supportive and others who harass her, Pacific Islanders living in a subsistence economy who are kind and helpful, some business people who recognize her talents and others who seem clueless.

Her book charts a course over five decades living through the challenges women faced as they moved into much more prominent roles in the professional world. She recounts the challenges and the factors that led to her successes. Her approach to writing the book is matter-of-fact and straightforward; she often lets events speak for themselves and provides some life lessons at the end of each chapter. In some respects, the book is an auto-ethnography, a cultural description of her own life, which itself is composed of her experiences in the diverse cultures and subcultural settings she encountered.

I knew Dr. Elizabeth Aden as "Betsy" in graduate school at the University of Pennsylvania, and I also did ethnographic fieldwork in a geographically remote society in the Southwest Pacific with many similarities to the society where she did her fieldwork. I find her descriptions of her experiences in graduate school in the 1970s, including her discussion of the sexual climate there, and many of her experiences living in a village in Melanesia, resonate closely with my experiences.

I recommend the book for its intimate view of a woman navigating challenging circumstances, its straightforward approach to that navigation, its acknowledgment of the help she received along the way, and its description of her continuing perseverance, resourcefulness, and successes.

—William W. Donner, PhD, Professor of Anthropology, Kutztown University; Freyberger Professor, Pennsylvania German Studies; and author of *Serious Nonsense: Groundhog Lodges, Versammlinge, and Pennsylvania German Heritage*

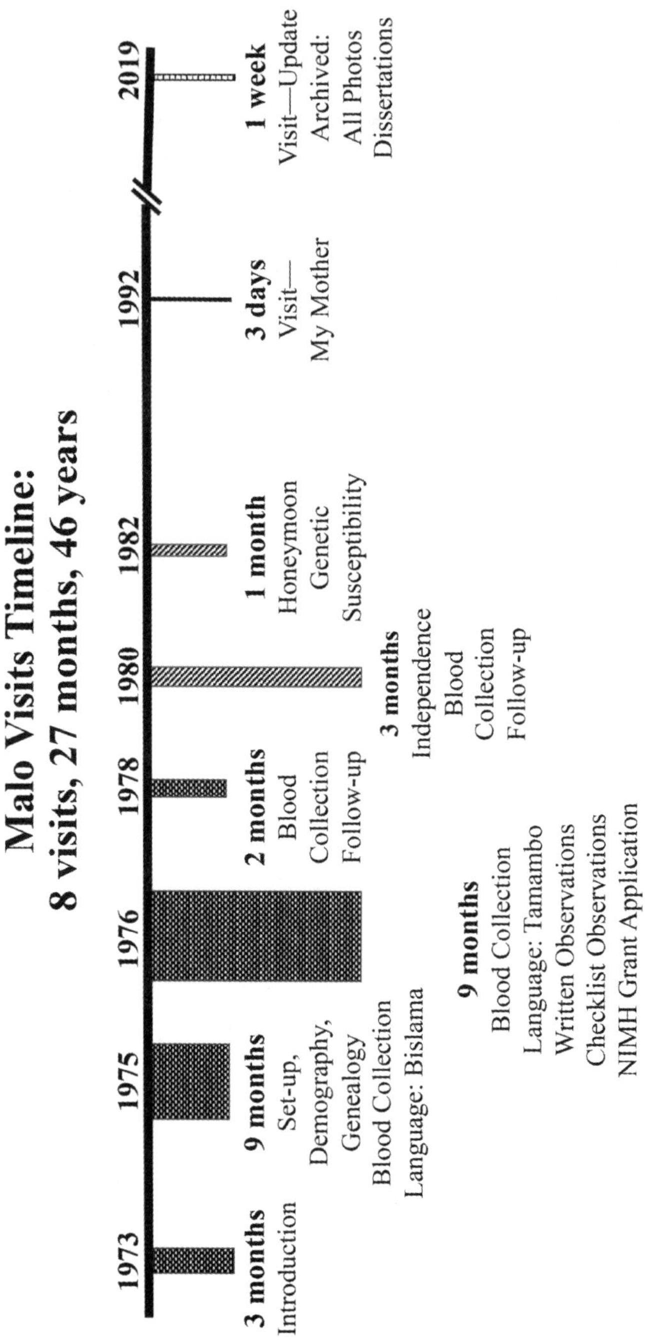

FIGURE 1: Timeline: Visits to Malo Island

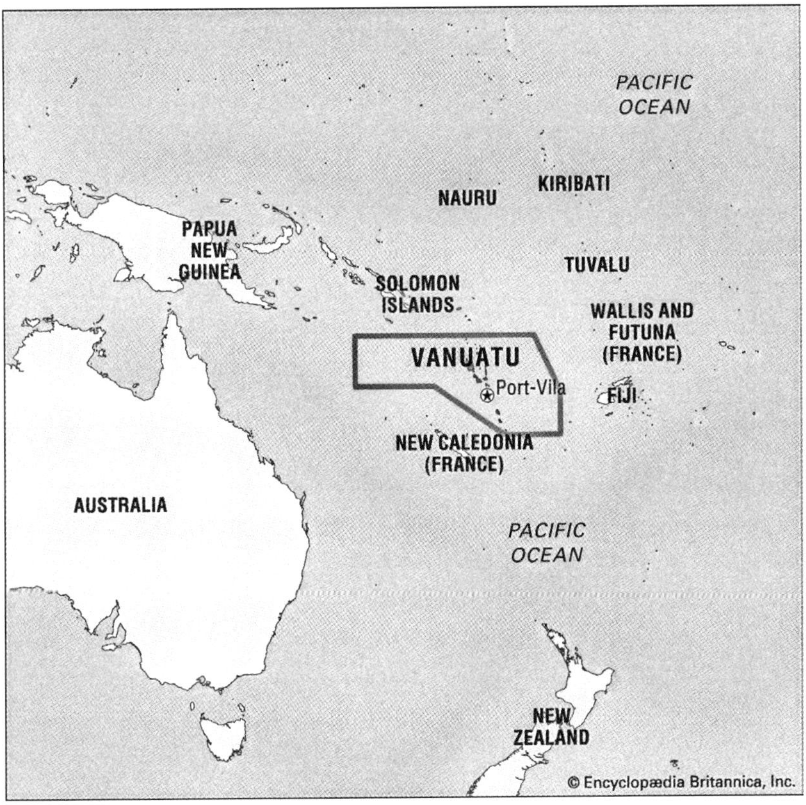

FIGURE 2: Location of Vanuatu (formerly New Hebrides), "Reprinted with permission from Encyclopaedia Britannica, © 2015 by Encyclopaedia Britannica, Inc.

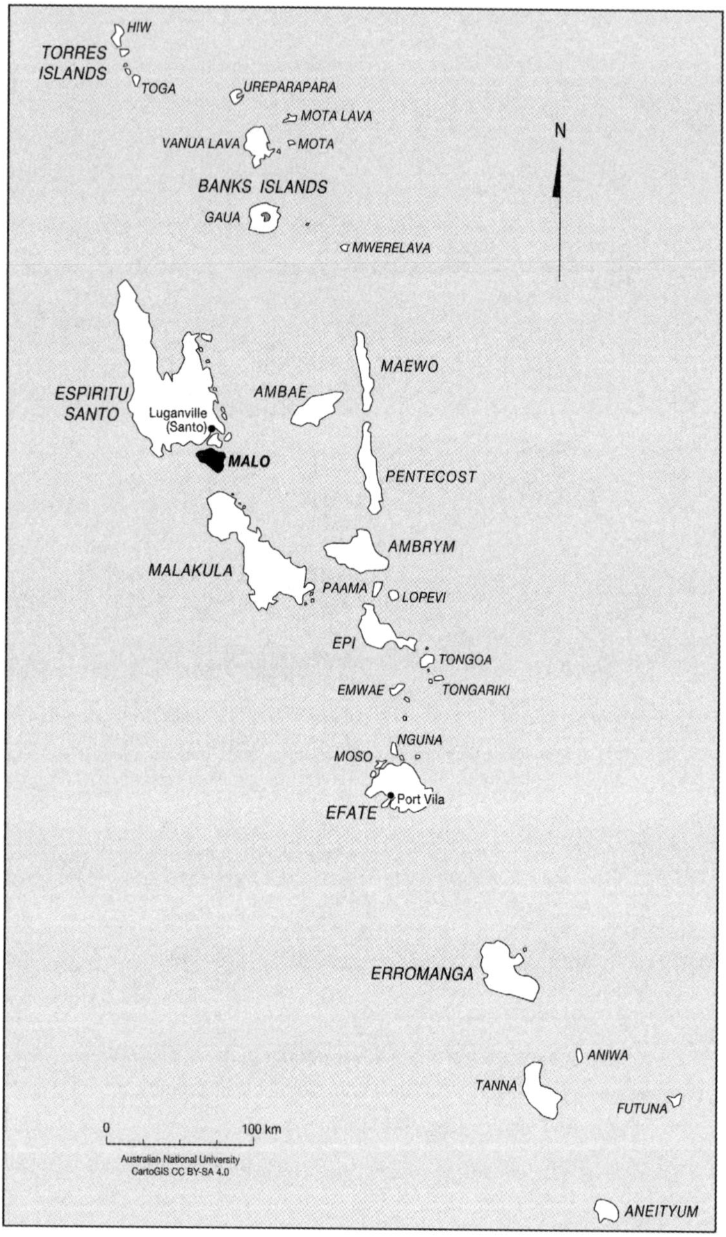

FIGURE 3: Location of Malo Island within the Vanuatu archipelago,* Map reproduced with permission from CartoGIS Services, Scholarly Information Services, The Australian National University.

* The island names are transcribed phonetically. The island Malekula can also be spelled Malakula.

Prologue:

Freedom

"Betsy, you have penis envy," Paul informed me, quite seriously, one evening. Throughout the early days of our marriage, he constantly told me I was doing something wrong or what I said was stupid. This was the last straw.

"Oh, please," I replied. "I am done with this!"

Penis envy was neither novel nor interesting. My mother, who was also a Freudian psychiatrist, had gratuitously psychoanalyzed me and my friends for the past ten years. I didn't need a husband doing exactly the same thing. At twenty-one, I wanted to go out with friends and have fun. I wanted life to be enjoyable. Furthermore, I thought life *should* be enjoyable. But I felt trapped by his expectations.

I'd married at nineteen because it seemed like a good idea. My mother was married at that age and had me two years later. Paul was cute, from a prominent family, and starting his internship—all in all, a good catch. Then we moved from Berkeley to Cincinnati, and he started his psychiatry residency and brought Freud home.

One evening, when all the Ohio universities were closed for the remainder of the spring semester after the Kent State massacre in 1970, we had Paul's fellow psychiatry residents over for dinner. During dessert, Paul took a telephone call. Once he left the room, I said, "I had the most amazing dream last night. It was so vivid and real. Would you like to hear about it?"

"Yeah, tell us," the residents enthused. Their eyes were fixated on me.

I chose that moment to savor another spoonful of ice cream and slowly withdrew the spoon from my mouth. "Snakes. I was asleep in my bed and then aroused from my slumber by these big gray-and-orange snakes coming out from under my bed and slithering . . ."

I had their attention. There is nothing Freudians love more than someone sharing their inner thoughts. I was making it all up, throwing in everything I had absorbed over the years and embellishing my "dream" with symbols I knew they would relish.

"Tell us more, Betsy!" they begged.

"Well, there were dragons, too. They were eating the snakes when out of the cave came . . ." I paused for effect. "Do you want to know more?"

Of course they did. I piled it on until Paul returned.

In the psychiatry department, divorce was a career-killing move for a young resident. The rationale was "If you can't control your wife, how can you counsel other people?" Divorce was thought of as a serious character defect. My "dream," shared with his colleagues, was my parting gift to him.

"She was totally out of it!" they'd say, and then add under their breath, "Definitely neurotic, maybe even psychotic." They would tell everyone that I was crazy, and what a relief it was for Paul to be rid of me.

A few days later, I returned to California. I found a lawyer who served Paul with divorce papers, but not until I had made sure I was financially secure. Just before we were married, I'd hit the jackpot when my great-aunt Maybelle died.

My maternal grandmother had been the only one of her siblings to have children. Her father had invested every spare cent he earned into buying stock in the fledgling company where he worked. The American Telephone and Telegraph Company's (AT&T) stock did well over the ensuing seventy years. On the death of his last surviving child, Maybelle, his portfolio was put in a trust to be shared equally between my mother's and my uncle's children.

On my twenty-first birthday, I received a quarter of my inheritance, including a lump sum of $10,000, which I considered a fortune.

"If you deposit that check into our joint account, it will prove you love me and are committed to our marriage," Paul had said.

"I love you, but my mother says it is my separate property," I replied.

"If you don't share it, I'll think that means you plan to leave me."

"I don't plan to leave you." So I had proved it by putting the money into our joint account.

Upon leaving Paul and arriving in San Francisco, I created a new bank account, in my name only, and transferred my inheritance into it. *Then* I called Paul, with twenty-four hundred miles between us, and courageously announced, “I’m getting a divorce.”

A day later, he called back. “I checked our joint account. You robbed it. You stole *ten thousand dollars*.” He was angry and called me a thief. If taking your own money back to escape a marriage that gave you nothing made you a thief, then, sure, that was me!

At twenty-one, I had the freedom and the resources to explore and see where life and opportunity would lead. I didn’t know where it would take me. I was ready to step into the unknown on my own.

Join me on this adventure.

MUD, MICROBES, & MEDICINE

Chapter One:

Culture Quakes (Santo and Penn)

From the air, Espiritu Santo, the largest island in New Hebrides (now Vanuatu), seemed like an endless deep-green mass, a wealth of different types of trees. Wooded and seemingly impenetrable, there were groves of coconut trees, banyan trees, mangrove swamps, and countless types of incredible, dense foliage.

The white landing strip was made from crushed coral, and as we approached, I could see people shooing cows away from the grass adjacent to the runway—for sure, I wasn't in the United States anymore! It was gloriously sunny. Outside of the terminal, casually dressed people waited for the six of us to disembark. There was no bustle at this tiny airport. A small concrete structure provided shelter if it was raining, and there was no baggage claim.

The last to depart the small twin-engine death trap, I stepped off the Islander and collected my stuff. My hosts were waiting for me on the airport apron. In my elegant khaki Safari Barbie ensemble, wearing my red backpack and clutching my brown canvas shoulder bag filled with tetracycline, I made my way to meet John and Karen. John was a graduate student at the University of Pennsylvania doing archaeology fieldwork on Malo Island. I feigned my best American-in-paradise persona but tripped as I walked toward the couple.

"Hi. It's so good to meet you!" I gushed. I studied Karen and then congratulated her. "You're pregnant!" I exclaimed. "How lovely!"

"No," she informed me. "I'm just fat."

I smiled my way through it but died inside. *Not a good start, Betsy, not a good start.*

The situation was defused when John suggested that we all go for lunch. We took a taxi into town to La Fief, a casual French restaurant. The first course served was a green salad.

"I'm not supposed to eat salad," I told them, thinking of the many health warnings I'd been given before my trip began—*don't eat uncooked vegetables, never touch unpeeled fruit, never drink the water, and remember, ice contains water.*

"Don't worry. It's fine," John replied. "You can eat everything—ignore the scaremongering. Trust me."

They were alive and looked healthy, so I did. All the rules went out the window with that first bite of crisp butter lettuce coated in oil and vinegar. I ate the salad; I ate everything that was put in front of me and waited for the diarrhea to arrive within the next twenty-four hours. It didn't—I had taken my cue from the "locals" and survived.

Culture Shock 101

How and why did I spend the summer of 1973 on a remote island in Melanesia, followed by seven additional trips over the following forty-six years? This die had been cast when I opened a letter from the University of Pennsylvania.

> *Dear Miss Dickie:*
>
> *I am happy to be able to inform you that we can offer you admission to the PhD program in anthropology in this department starting in September 1971 without financial aid.*

The eldest of four children, I am the sole daughter of my parents, each of whom was married four times. My three brothers and I span twelve years, have three different fathers, and grew up in two separate homes, but we share a common stepfather. Complicated? I've just begun. My maternal lineage is graced with high achievers. Great-grandmother Mary won the title of Checker Champion of Boston. Her daughter Minnie passed the Massachusetts Bar in 1921.[1] Minnie's daughter Marilyn, my mother, was one

1. Minnie, a member of Portia School of Law's first class, graduated in 1920 with a bachelor of laws (LLB) degree. The school was founded in 1908 as a law school for women. It was reincorporated in 1918 as Portia School of Law and was granted the ability to award the LLB degree in 1919.

of three women to graduate from the University of California–San Francisco (UCSF) School of Medicine in 1956. Then there is me. Grad school was a long shot.

For the most part, I had done what was expected of me—but just barely. An underachiever, I applied to seven graduate programs hoping I might get into at least one of them. Penn was among those schools because my teacher at the University of California, Berkeley, had taken a professorship there. I knew he would remember me, since I'd earned an A+ in his fossil primates course. I hoped the University of Pennsylvania, which I thought was another state school, would take me with my mediocre grades and worse test scores. When they did, I accepted, and the adventure of my life began.

I remember my first day very well. Dressed in my UC Berkeley best—a tie-dyed T-shirt, my mother's blue jeans from Wellesley, navy wooden clogs, and my hair in pigtails with daisies poking out from behind my ears—I clomped down the red brick path toward a group of well-dressed twentysomethings who were milling around the steps to a medieval-styled, metal-laced wooden door with discreet signage: ANTHROPOLOGY DEPARTMENT. I had arrived.

The door opened to reveal a short, neatly groomed man wearing a beige summer suit. He introduced himself as the department chair and then invited my class of twenty doctoral students into the building. As we filed in, he looked at me and said, "You must be the *one* we took from Berkeley."

At that moment, I realized how different I appeared. My male classmates were in khakis, polo shirts, and deck shoes, and the women wore shifts. I was informed there was a dress code, literally. After a letter to Mom, dresses from home soon arrived.

The conversations and language were also unfamiliar. "Where were you prepped?" people asked each other. "Saint Paul's" and "Choate" were common answers. To me, an East Coast neophyte, it seemed like a lot of people had undergone surgery—the only context I knew of where people were "prepped."

Other things were different, too. As our class gathered to go to a Chinese restaurant, one of my new colleagues asked, "Where's Alan? Does anyone know where Alan is?"

"Which one is Alan? What does he look like?" I asked.

"He's the Jewish-looking guy with red hair." Huh? Red hair was the only part of the response that I understood. It was more new information.

I mentioned this exchange to my British archaeology professor, with whom I had established an engaging repartee. His response: "That's nothing. In Ireland, your life can depend on differentiating 'Catholic skin' from 'Protestant skin.'"

In 1971, the East Coast had a very different culture from the West Coast, where a person's religion wasn't significant and certainly not a descriptor. I was experiencing culture shock.

It took me six months to realize I had inadvertently landed in an Ivy League school and one of the best anthropology departments in the country.

Barbie on Safari

One of the key decisions for anthropologists is where they will do their fieldwork. The decision focuses your attention and coursework on that specific region. It can also determine who will be your adviser. Where in the world did I want to live for two years? Alaska? Too cold. The Middle East? Too hot. How about the South Pacific? I'd analyzed Polynesian languages for my master's thesis because I had a crush on my professor.[2] When he offered me the word lists he had collected while doing his fieldwork in West-Central Polynesia, I said yes. I'd started Penn thinking I would study human paleontology, but I quickly realized my dissertation would consist of analyzing fragments of fossil bones, likely in Africa—that wasn't so appealing. Polynesia, however, could be a desirable place for fieldwork: Those islands had beaches and—if done right—French food.

There I had it: I was making a major life decision based on personal desires and fun. Good food, sunshine, and beaches—why not decide your life on that if you could? Somehow, a dissertation would emerge, but in what way or on what topic, I had no idea. *Well, I'll cross that bridge later*, I told myself.

I was starting my second year of grad school at Penn. I had been coding the Trukese-English Dictionary for almost a year as an assistant to Ward Goodenough, who became a trusted mentor and the first of my thesis advisers.[3] He was a well-known cultural anthropologist whose geographic expertise covered

---

2. My master's thesis used linguistic analysis to determine whether the language in the northern and southern Ellice Islands (now Tuvalu) came from the same or multiple sources. Elizabeth R. Dickie, "A Lexicostatistical Analysis of Six Polynesian Languages: A Study in Methods" (master's thesis, University of Cincinnati, 1976).

3. A Trukic language of the Austronesian language family spoken in the Caroline Islands in Micronesia.

the three regions of Oceania.[4] Goodenough was a short, stout man whose face was dominated by large, dark-framed glasses that magnified his bright-blue eyes. He had a full head of thick, wavy brown hair with enough gray to look wise. I was awed and intimidated. He was serious and conscientious while he showed me what he wanted. Once he realized I could and would do a good job, he relaxed and became a font of encouragement and support. He boasted about buying his Ivy League Harris tweed sports coats at the Bryn Mawr thrift shop. He talked about how different customs were on Truk, an island in Micronesia where he'd conducted fieldwork over a twenty-year period.[5]

When I laid Helen Gurley Brown's best-selling book on my desk, he said, "*Sex and the Single Girl* is nothing new." I was intrigued and asked him whether there was premarital sex on Truk.

"When I first arrived, in the late forties, I was surprised to learn there was no stigma attached to premarital sex. Even virginity wasn't valued on Truk," he said.

"How do you know?" I asked.

"There is no privacy in their houses, so I asked about how, when, and where couples had sex. That opened the door. Children are aware that their parents have intercourse their entire lives and, when they reach puberty, they experiment and talk with their friends. It is very open and not taboo. In general, premarital sex is discreet and prohibited between close relatives. After marriage, adultery is frowned upon, with the exception that a man is free to have sex with his wife's sister and his brother's wife."

"How does that work?"

"Truk is both matrilineal and matrilocal, so the husband's genes are staying in the woman's family, as it were. It is a handy system if either his wife or his brother is infertile," he explained.

"That makes sense," I said, and seized the opportunity to advance my agenda. "Wouldn't it be a good idea if I went to the South Pacific and got a sense of what fieldwork might be like?"

I was really thinking, *Beaches for the summer!* I imagined myself swimming in the warm blue ocean and watching the waves break while walking or basking on the sand. It would be a paid vacation in heaven.

---

4. Oceania, or the South Pacific, is divided into Melanesia, which includes the islands between New Guinea and Fiji; Polynesia, which includes the islands that fall within the triangle anchored by New Zealand, Hawaii, and Easter Island; and Micronesia, which includes the islands north of the equator and east of the Philippines.

5. Now called Chuuk.

FIGURE 4: Betsy at work on the Trukese-English dictionary

"That's a very good idea," he said, likely seeing right through me. "Penn is administering a Ford Foundation grant to support anthropology fieldwork. Since your master's thesis examined migration patterns in Polynesia, we can rationalize the trip as reconnaissance for future fieldwork in Oceania. I know the perfect place. You can go to Malo Island, where one of my doctoral students is doing his research in archaeology," he continued. "John's expanding on his master's thesis. He found a Lapita site on Malo Island and is doing more research on the culture of these early settlers in Melanesia. He and his family are living on Malo, an island in the New Hebrides. You can spend the summer with them and get a sense of the island's culture. It'll be a great introduction to fieldwork for you."

It wasn't Polynesia, but it was in Oceania. It was a start. I got the grant and told my mother I was going out to the field. Christmas was approaching, and she used all her gift-giving imagination to select an outfit appropriate for a young woman venturing into the jungles of Oceania. There was only one place to shop for it: Abercrombie & Fitch, which was then an upscale sporting goods store.

I unwrapped the box and saw the A&F label. Gently, I peeled the white tissue paper apart. Behold—the A&F box contained a khaki-colored ensemble that could have been labeled "Safari Barbie Goes to the New Hebrides." I lifted the first piece, a vest with a sheepskin patch for fishhooks and a leather patch to secure the butt of a rifle. *Hmm*, I thought, *I'll be doing fieldwork, not going on safari.* The next piece was a long-sleeved shirt—that could be useful, but I was going to the tropics. The last piece was a culotte skirt with pockets—something I would actually wear. Each piece of the khaki suit was labeled "Dry-clean only."

I'm not sure how many dry cleaners Mom thought there would be on the island, but I was reasonably certain that number would be zero. I was ready to become the exotic fieldwork poster girl for one of A&F's nonexistent ad campaigns—*thanks, Mom.* The outfit ensured that I would still look feminine while undertaking my serious work. I appreciated her effort for me to have the best, even if the best was impractical.

Mom was a physician who often overdid things. Before I left for the New Hebrides, she decided I needed to speak to a tropical medicine expert at UCSF.

"You could be a savior, an absolute angel of mercy to the 'savages' you will encounter," he told me.

"I believe the word you're looking for is 'people,'" I countered.

"You need to take as much tetracycline as you can carry," he continued. "And antimalarials." He was the one who had warned me off everything from fruits and vegetables to water.

His advice boiled down to this: Don't eat or drink, and take all the antimicrobial medicines in the Western world. I walked out of his office with prescriptions that became fifteen pounds of antibiotics stuffed into a brown canvas shoulder bag. So, when I started my sojourn as Safari Barbie, I was secure in the knowledge that these pills would save all the natives' lives. It was his vision, not mine. He believed I was going on a life-threatening mission into the wild, whereas I was thinking, *Bain de Soleil.*

I was able to ignore the disconnect between our visions until I went to the library, where medical journals described the different ways ocean creatures could kill me. I knew there were no deadly animals on the island, but I had no idea that in the sea lurked many terrors, worst of all, sea snakes. *Sea snakes.* If one bit me, I would have about an hour to live. My problem was that I had 20/400 vision. A sea snake could sneak right up to me; I would never know, since I swam without glasses. I might only see it when it bit me. I was terrified.

My fantasy was still swimming at a nice beach with palm trees (well, perhaps I'd dropped swimming from the fantasy by this point) and living with a nice family in a delightful French–English colony (see Color Plate 1: Champagne Beach and Hat Island [Espiritu Santo]). All very civilized. Sure, there was malaria, but I had pills. Survival for two to three months in the New Hebrides would be easy; people lived and survived. On the other hand, experts said death was everywhere from swimming in the ocean to eating the food, contracting infections, and being bitten by insects. I had no way to reconcile these two extremes until I arrived in Santo.

Living History

After surviving my flubbed arrival at the airport and my disregard for dietary precautions during lunch, I relaxed.

"Soon you'll meet Ken Hutton. He has an empty apartment where we'll be staying for tonight," John announced. "Ken's Australian and the dentist in town. He's a great friend. I met him when I was here working on my master's. I'll give you a tour on the way to his house."

As we walked down the main street of Santo, John explained how the road had been made.

"In World War II, when the Seabees arrived, they plowed this main road. They then dredged up live coral, put it on the road, and ground it down flat with steamrollers. Next, they watered the road so the coral would grow together. They repeated this a couple more times, and the result is this incredibly durable, waterproof, pothole-free coralline road."

"Impressive. It also explains why the road is so white. Does it reflect the heat?"

"Yes. So it's even hotter if you walk on the street. It is best to stay in the shade of the covered wooden sidewalk. How are you doing?"

"Fine," I lied. My back was drenched with sweat from wearing my back-pack. It was hot. I was glad John was carrying the drug-filled tetracycline (tetra)-sack. As we walked in the partial shade of the covered boardwalk next to the stores, I watched in awe as the sweat-free local men passed us in their flip-flops, T-shirts, and Bermuda shorts.

"I feel like I'm in an Old West town, only the horses have been replaced by sedans, and instead of saloons, we're passing storefronts offering Chinese goods."

"Don't worry, there are plenty of bars here. Mao's is the main bar, especially for expats," John said, pointing diagonally across the street to a nondescript brown wooden building. "Just before you get there is the Tropicana, which

FIGURE 5: Main road in Santo Town (Luganville, 1973)

makes the best French bread and pastries." He then stopped for a moment and turned around. "See that big blue Quonset hut with the CFNH on it?[6] That's the French supermarket. Next to it is the butcher shop, and behind it is Chez Pappy's. Pappy's son, Michel, is the butcher. He sells Charolais beef that melts in your mouth. We eat well here."

"I have a lot to look forward to," I said, rubbing my stomach and thinking that I might have lucked out after all. "But where did that Quonset hut come from?"

"Santo, or Luganville as the French call it, exists only because of World War II. The area between Santo and the adjacent islands forms one of the largest protected natural harbors in the Pacific. It was also close enough to the Pacific Theater to be the staging base for the Battle of the Coral Sea, yet far enough to offer the troops R & R. They say over one hundred thousand troops came through here. The Americans built the town and the three landing strips on the island. Seems everyone pitched in: the Seabees, the navy, and even the army."

We turned around and kept walking. Soon we were off the boardwalk and on the shoulder where the coralline road met the packed dirt. Every so often, a jeep or truck passed us.

"See that jeep?" John asked.

"The small one without a roof? Looks like it's from *Patton*?"

"That's the one. It is a jeep from the war that was shipped over in fifty-five-gallon drums filled with oil. It was assembled here and, for a little money, you can buy one still in the can."

6. Companie Française de Nouvelles-Hébrides.

"Canned jeep? That's a new one." I tried to visualize how the pieces would all come together. John pointed out other trucks from the war as well.

Santo was the island's hub and much more developed than where I was ultimately headed. It had a reasonably sophisticated expat community of fifty to seventy workers, who, with their spouses and families, frequented the stores, restaurants, and bars. There was also Burns Philp, an Australian supermarket, to complement the French CFNH.

John explained that the working expats went for long lunches at the handful of bars and restaurants. On the weekends, they drove to the Santo Golf and Country Club ten kilometers away in Palikulo Bay. Life was simple for the expat community during the week—the men worked and imbibed throughout the day, and the women stayed home to shop, cook, and take care of the children.

I had stepped back in time, or onto a movie set from the 1950s. The one-street town with a laid-back air was a living anachronism. I wasn't in the 1970s anymore; I was within touching distance of World War II.

We reached a fork in the road. "The Sarakata River, where I tie my boat, is to the left and Ken's house is to the right," John said. We walked a block to the right and stopped at a large single-story, white-washed concrete house. On the left side of the building was a brown-painted door with one word: DENTIST. Ken's duplex was our base in Santo.

FIGURE 6: World War II Quonset huts in Santo (Luganville) (1973)

Later, when I walked next to the Sarakata River at the edge of town, I would sing a refrain from one of the songs in *South Pacific* about leaving America and coming to a place with coconut palms, banyan trees, and coral sands—the "Tonkinese" was the only missing piece.[7] By tapping into my childhood memories of the movie and the songs I'd long ago memorized, Michener's work made Santo more accessible to me (see Color Plate 2: *Tales of the South Pacific* by James A. Michener).

John lent me an old moldy copy of *Tales of the South Pacific* that was among Ken Hutton's books. The inside cover was stamped "Vanikoro Library" and had come from the nearby Solomon Islands. Perfect. I read it and learned more about the war experience in the Pacific Theater. In Santo, where Michener had been stationed, I met Madam Ratard, who some thought inspired the memorable character Bloody Mary. The French plantation owner on whom Emile de Becque was based had died during the war. There was a decaying sign on the island of Aore that had once read American Beach; this beach was for nurses like Nellie Forbush. Bali Hai, the mystical island on the horizon penned by Michener, was the island of Aoba forty miles to the east of Santo in the other arm of the "Y" that gives the archipelago its distinctive shape.[8]

And because World War II was still alive in people's thoughts (especially the men's), I heard a lot of stories that brought history to life. Sometimes I would wander down to the Mission House and talk with the older men in Bislama (pidgin English).

"Tell me about the war and the soldiers you met," I'd say.

"Planes would fly over Malo," they would tell me. "Lots of planes. They made a lot of noise."

Or, "Sometimes soldiers would come over to Malo. We fed them *laplap* and they gave us gifts.[9] This shirt is a gift from a man named Fred. I traded a handwoven bag for the shirt. The shirt is very good. My brother got a tray.[10] He beat the tray until it was flat and could hold *laplap*."

"Ships were everywhere. There were so many ships. They come, they go, they come for a long time."

---

7. Tonkinese refers to Indochinese, including Vietnamese indentured servants, who were brought to New Hebrides to work on the French plantations from the 1920s until the beginning of World War II. At one point, 10 percent of the population was Vietnamese.

8. Aoba is now called Ambae.

9. *Laplap* is a baked pie made of grated yams and often topped with coconut cream.

10. A steel mess tray divided to separate different elements of the meal, i.e., vegetables, meat, and dessert, etc.

"The Americans who come were good. They were nice. They brought gifts."

The islanders remembered those times fondly, never having experienced battle like the nearby Solomon Islands.

The islanders' vision of Americans was very positive, a big plus for me. They thought of us as kind and generous. The Americans had given the islanders things they could spare, such as shirts, caps, gas masks, metal cups, mess trays, and utensils. These gifts—and the associated memories—were cherished and on display whenever there was an important event.

One day, when I accompanied John on a shopping excursion in Santo, I decided to spend some time getting to know Ken. Ken had arrived in Santo well after the Americans had left. As far as I knew, no Americans had chosen to live in Santo after the war.

"What was Santo like when you arrived?" I asked.

"Pretty much what you see today. You Septic Tanks turned the place from a plantation into a town," he said in his clipped, Cockney-esque Australian accent.

"Septic Tank?"

"We Aussies like to rhyme, and we love nicknames. You're both a Yank and a sheila."

"Sheila means 'woman'?"

"You're bloody smart, Septic Tank," was his retort. "You want to know more about the war? Let's visit Million Dollar Point."

"That sounds promising. Let's go," I said.

We got into his white Toyota sedan. We took a back road through streets lined first with concrete houses and then wooden shacks. The road had not been sealed; it was rutted, and Ken drove slowly over and around the bumps. Ten minutes into the drive, the road ended. We got out of the car and walked to the edge of a desolate fringing coral reef. At the water's edge, I stopped to look more closely at a large block of rusted metal.

"Take a good look at it," Ken said. "You might be surprised."

I squatted and examined what was now obviously a machine of some sort. Closer scrutiny revealed the name "Friden" and corroded keys with an occasional number. "It's an adding machine."

"That's right. Now look over here," Ken said, pointing to a much larger object farther out on the reef. "That's a bulldozer."

"What happened here?"

"Simple. When the war ended, you Septic Tanks offered to sell the equipment you brought to Santo. You weren't going to take it home, and the Brits and Frogs knew that. They thought, *Why pay anything when it's going to be free once the Yanks leave?* So the *colons* informally divided up the would-be spoils of war and decided to wait it out. But you Septic Tanks weren't having any of that. You decided that if you weren't going to be paid for it, you would drive everything into the sea . . . construction equipment, office machines, chairs, you name it. Yankee efficiency. That's why it's called Million Dollar Point. There's millions of dollars' worth of machines and equipment that was deliberately ruined."

We walked on the coral, stepping around or over the rusted relics.

"How did they get the equipment into the water?" I asked.

"You Yanks built a ramp out to about a quarter of a mile. Then everything was driven into the sea. Hundreds of tons, everything from Coke bottles to tanks. Some of it is under a hundred feet of ocean and some's washed up on the shore."

The scope of what had been driven into the sea was stunning and brought home the inanity of war—it reinforced what I'd experienced during my college years when protests against the Vietnam War roiled Berkeley and other college campuses.

Today, Million Dollar Point is a tourist and diving attraction.

I felt a little like a tourist myself; after all, I was on the first overseas trip I'd taken since I was three years old. Everything I saw seemed new and different. Even the bathrooms were different, with the sinks and toilets in separate rooms. I thought they should be together—it was basic hygiene! The toilets flushed differently, too. This was only the beginning of the adjustments I had to make.

I searched for the familiar but found only the foreign. Santo had suffered a big earthquake the year before I arrived. I could still see its destructive effects—huge fissures in concrete foundations, walls with spiderweb cracks, rubble piles from collapsed buildings—and it was much worse than anything I'd seen in San Francisco. I was definitely a stranger in a strange land. I had new coping skills to learn, but I noticed that the people here, both natives and foreigners like myself, looked happy and relaxed. I just had to learn what they did and how they did it.

Adapting was easy on days when the sun was out—the contrast of emerald-green trees and sapphire ocean against the blue sky was spectacular. When the sky was gray, the ocean was gray, and the foliage muted to olive;

sea, earth, and sky blended into each other. The days of deep green and blue kept me upbeat while the days of gray could be glum.

Overall, I adjusted to the semitropical climate and found it quite pleasant. The islands were neither humid nor hot, and my aluminum-lined sleeping bag (which I had brought despite being advised not to) kept me plenty warm.

I was unprepared for what I encountered in Santo during those first two days. I had no idea of what to expect on Malo. I was hours away from finding out.

Lessons Learned

- Ask and apply for what you want; you just might get it.
- Important decisions can be made based on seemingly frivolous criteria, like having fun.
- What you don't know can kill you.
- You can survive, even thrive, by eating as the locals do.
- Childhood memories can provide comfort in a new and strange place.
- The American impact from World War II endures.
- Excessive greed gains nothing. Best case, it results in a tourist site.

Chapter Two:

M-Day—Landing on Avunatari Beach (Malo)

"Are you ready to go and see your summer home?" John asked the next day. "Let's load all the food and supplies into Ken's car. Karen will go with him, but you and I'll walk the half mile to the boat."

We walked along a treeless road in the midday heat down to the Sarakata River, which, fortunately, was lined by trees that gave relief from the relentless sun. The river's edge was defined by a partially corroded steel seawall that had been installed over thirty years earlier. The retaining wall was still intact and kept the river from eroding its banks while providing a convenient docking area. John's boat was tied there.

"See the island in front of us? That's Aore. See the island behind it, in the distance, the one that looks like a hat?" John asked, pointing. "That's Malo."

"The island with a low, flat brim and a flat crown, like a broadbrim woman's straw sun hat?" I asked.

"That's the one. See the point on the far right? That's where we're going."

"How far away is it?"

"About ten miles. Look how protected it is. The passage between Aore and Espiritu Santo forms a large canal, and it's where the Pacific Fleet was harbored."

"Which is why everyone refers to it as 'the Canal,'" I said, reinforcing the now obvious in my mind.

"Just beyond Aore is the open sea." John finished loading the boat, placing my backpack and food under the small covered part of the bow. "Hop in.

Why don't you sit up front near your things? You'll stay drier."

With my tetra-sack resting on my back, I stepped down onto a bench in the open-hulled ten-foot aluminum fishing boat. John extended his hand. He then helped Karen into the boat and took his captain's seat next to the outboard motor.

I kept a tight hold on my precious shoulder bag, which was filled with the island's salvation (antibiotics—my tetra-sack), as we sped through the Canal. When we left the shelter of Aore, the water was choppy and the ride bumpy and wet. We were crossing deep water in a part of the Pacific Ocean where tsunamis are relatively common. The fifty-minute ride was both terrifying and exhilarating. I felt vulnerable because the trip was rough in the small boat, but I was fascinated by the large, hat-shaped, dark-green island we were approaching. I also appreciated that the speed kept me from seeing the sea snakes and sharks that were surely lurking below.

I didn't see a dock as we approached Malo. John slowed the boat to a crawl and navigated the fringing reef, a seemingly impenetrable barrier of sub-surface coral. Only a few tiny passages allowed the boat to approach the shore. Two islanders appeared as if on cue and waded out to help beach it. About five feet from the shore, they stopped our drifting speedboat. One man uttered something unintelligible and picked up my precious red backpack and tetra-sack. He tossed my belongings to another man on the shore and came back to carry the boxes of supplies John had purchased in Santo.

These men walked effortlessly between the boat and the shore. I was confident I could do the same. I swung my legs over the side of the boat and slid into the water in my tennis shoes. To my surprise, I could not find secure footing and stumbled. I desperately grabbed the side of the boat to steady myself. I stood paralyzed in the tepid, knee-deep salt water until the man who had transported my belongings to shore offered me his arm. Even with his help, I looked like a drunken sailor as I stumbled on the uneven coral. I watched in amazement as my guide walked barefoot on the sharp coral, the soles of his feet thick and calloused. He was confident and graceful, while I struggled and slipped. I made it to the beach, relieved to be on *tierra seca*.

That was my introduction to Malo. Within seconds, I realized how unprepared I was to live there. I looked and felt so naïve. I'll never know whether the islanders were laughing at me, because they offered me only kindness and consideration.

In time, I would be able to embark and disembark in flip-flops, but I always had to watch where I stepped. I was never able to walk securely across the reef as the islanders did.

FIGURE 7: Malo Island viewed from the Sarakata River

FIGURE 8: Avunatari Village (Malo Island) speedboat landing

On the Malo beach, I was introduced to some of the islanders, one of whom carried my backpack and talked in Bislama. I smiled and looked for nonverbal cues. John chatted in Bislama with the group carrying our things. I caught a few words, such as "conch," "snail," "rockfish," and the phrase "*man em he die*." I asked John whether I'd gotten the gist of the conversation and what I'd missed.

"Just be careful where you touch on the reef. The conches are obvious, but I'd avoid picking up any pretty cone-shaped shells to get a closer look. The spotted ones are deadly."

"But isn't the reef filled with shells?" I looked down at the ocean's edge and confirmed my observation. "Weren't the locals stepping on shells when they carried our things?"

"Yes, but these shells aren't alive," he replied. "The part of the reef that's awash is fine. It's only when you get into the slightly deeper water that you have to be careful. The worst and hardest to spot is the stonefish."

"Because . . . ?"

"Of their exquisite camouflage. They live in the reef and are indistinguishable from the coral. Their spines are poisonous and can kill you as well. If you're lucky, you'd get a single spine prick. That would be painful and you'd have to go to the hospital. My recommendation—don't step on one."

Great. More animals in the sea to kill me. Maybe my fantasies about beaches need some updating, I thought.

We walked up from the beach, stepping over mangrove roots into a grass-covered clearing the size of two football fields. In the middle of the clearing, built on a concrete slab, was a yellowed wooden building with a veranda and a pitched, faded red corrugated tin roof. Beside it was a frame, made from tree trunks, that held a bell. There was a concrete building nearby.

"Welcome to Avunatari Village. In case you're wondering, *avunatari* is a type of breadfruit," John announced.

I was confused. My mental image of a village included dwellings grouped together. "Where are the houses? Where are the villagers? All I see are these two buildings."

"This is more of a meeting place; people live on their traditional lands. On Sundays, islanders come here to the Mission House for church or for social events. That concrete building is the schoolhouse. It's the British school, and it's free. It goes from first through sixth grade. Somehow, all the different grades manage to fit into that one building. There is a French school, but it's on the other side of the island and has tuition. The advantage is that it covers the standard French curriculum."

"Do kids go on to secondary school? Or does formal education stop at sixth grade?" I asked.

"Most children in the British school stop, but a small percentage continue on to high school and college. Our landlord's daughter, Grace, is one of them. It's hard on the family because the schools are on different islands, with housing, tuition, and travel costs. I'm sponsoring Grace as a way of saying 'thanks' to Jole for all his help. One benefit offered by the French school is access to French universities."

John's explanation was helpful and whet my desire to learn more. The islanders cared about education and wanted to give their children opportunities. Their emphasis on education seemed unusual, and education led to important positions within the government, the major employer for islanders. Those children who went on to higher education were usually fluent in both French and English, which helped them secure employment rather than making a living with subsistence farming.

We walked farther up the hill and stopped in front of the Mission House. Lying in front of the building were two weathered marble tombstones. "Those are the graves of the first missionaries' children," John said. "They both died of dysentery in 1889. It's a shame. Drinking lots of water with salt might have saved them from death by dehydration."

We walked over to a tall headstone that stood by itself in front of the church.

"Look at this grave. See the name? Antas?" John asked.

I nodded and looked at the limestone monument engraved with dates and unfamiliar words.

"Antas was the missionaries' first convert and the grandfather of my landlord, Jole. He died in 1897, ten years after the first missionaries arrived." (See Color Plate 3: Antas's grave.)

My anthropology training had made me skeptical of missionaries because they wanted to change the Indigenous culture. I thought of it as a negative that Antas had been converted, and I wished that the missionaries had taken the hint and returned home after their children's deaths.

We walked past the Mission House, and John waved to some men talking on the far side of the open field. They sat by bamboo huts that looked as though they would not survive a single puff from the Big Bad Wolf, much less a tropical cyclone.

"See the houses over there where the men are sitting?" John asked. "These families are on their traditional land. The patriarch in each family donated land to the church and the school." On Malo, land ownership and residence were passed from father to sons.[1]

Women sat together farther down the hill, watching children playing on the grass in front of a row of houses. Daughters, when they married, lived with their husband's family. Family membership was traced through the father's lineage.

This is like the social organization I read about in my anthropology books. I am seeing it in action, I thought. It was my first anthropological observation. I was on my way!

We left the village and the open space behind. Ahead of us was a green forest shading a single-lane dirt road. We walked past a corrugated metal shack with the word "Co-op" painted in blue on a worn brown plank nailed above the door. All the houses we passed were made of split and woven bamboo walls with pandanus-thatched roofs. It was another world. I tried to absorb it all.

"The houses don't look very secure," I said.

"They aren't. There isn't a lock on any of them. Malo's culture is built on trust. In fact, the Europeans described the islanders as ignorant because they 'didn't even know how to lie.' Can you imagine that?"

I couldn't.

Counterintuitively, the openness gave me a sense of security. It reminded me of my childhood of unlocked doors and playmates just a knock away. The spontaneous smiles of the people we passed gave me a sense of comfort. Going between houses was simple; you either walked across the grass or followed a path. Meeting people was casual and uncomplicated, and everyone was friendly.

We continued on our way to John's house, which was located on the property of Jole Antas. As we passed, a group of older children greeted us: "Hello. How are you today?"

John replied, "Fine, thank you. And how are you?"

"Very well, thank you. Goodbye."

"That was a typical conversation in English," John said. I later learned it was more a catechism than a conversation.

---

1. The culture was both patrilineal (inheritance passed through the father's line) and patrilocal (people lived on their father's land).

We left the main road and walked toward John and Karen's home. We went through a barbed-wire cattle gate onto a single track through light brush.

"We're entering Avnaro, the home and land of Jole Antas," John said. John had met Jole when he accompanied his master's thesis adviser, archaeologist Betty Shutler, to scout for sites to excavate. They found a Lapita site, which John excavated that summer.[2] Jole was the only person he met who spoke English fluently, partly due to his medical studies in Fiji. He would later go on to become a member of Parliament.[3]

"Remember the grave we saw with the Antas name?" John asked.

I nodded.

"Antas was Jole's grandfather. The usual custom is for a child to take the father's first name as their last name. Jole broke that tradition and took Antas as his last name, and he's given the Antas name to his children as well. They are now following the American pattern. In contrast, there is Vira, whom you met at the beach. His father was Joseph, so Vira's name is Vira Joseph. Vira's son, Joseph, is Joseph Vira."

"So the father's name is generally preserved for a single generation?" I asked.

"That's right."

"It must make it hard to keep track of who's who after a couple of generations," I said.

"Not really. The men stay on the land of their male ancestors, and oral tradition keeps the names and lineage relevant, especially if there's a dispute."

In the clearing, I saw two clusters of buildings. The nearer compound looked newer and had features that were painted bright pink.

"The pink complex is ours," Karen said. "I picked the color."

It complements my Safari Barbie outfit, I thought. Their compound comprised four buildings, each made of thatch and bamboo.

---

2. Lapita pottery is very similar to pottery found in Luzon, Philippines. The pottery is associated with the Neolithic migration of Austronesian speakers primarily from Taiwan and the Philippines into Melanesia, Polynesia, and Eastern Micronesia. The migration occurred around 3,600 to 2,500 years ago (1600 to 500 BCE). The Lapita pottery found on Malo was dated to between 70 BCE and 1010 CE. See John D. Hedrick, "Lapita Style Pottery from Malo Island," *Journal of the Polynesian Society* 80, no. 1 (March 1971): 5–19.

3. Jole Antas was elected to serve as a member of the Vanuatu Parliament after the country became independent in 1980.

Home Sweet Home

"Let me show you around," Karen said, and opened the wooden door. "This is the kitchen. The floor is dead coral, which means we don't have to clean it. Any scraps that drop there are eaten by bugs living in the coral. John had the house built with bamboo walls only halfway up and screened in the rest. We're the only kitchen on the island with windows." Turning to face the back of the kitchen, Karen continued, "We have all the comforts of home, including a four-burner propane stove with oven, an electric refrigerator, a washbasin, and shelves for dishes and food."

"What about water?" I asked.

"See that big tank outside?" John said. "It's a two-hundred-gallon collecting tank. We collect rainwater from the kitchen's slanted corrugated metal roof. I put a fine mesh screen over the hole. The water's been fine and tastes great. I'll show you." He took a glass down from the shelf.

We walked through the back door to the tank. "See, it's just like in the States. Turn the spigot and you get water. Try it." He handed me the filled glass.

"It tastes fine," I assured him. This was our fresh water source—our "running water." It was like home, with a difference. If you weren't careful, it could run out, especially if it didn't rain.

"Next on our tour is the bathroom." John walked over to a small building to the left of the water tank. He opened the door and revealed a covered pit toilet crowned with a hot-pink wooden box seat and a strategically placed hole covered by a black toilet seat. A roll of pink toilet paper was perched on a heavy-duty nail.

"Nice," was all I could say.

"Be sure to sprinkle ash down the hole every time. It helps keep the flies to a minimum." He pointed to a wooden box filled with ash and an empty tin can. "Next is the shower." Attached to the outhouse was a bamboo bathing stall. "This is a shower bucket." John took it down from its hook to demonstrate. "It's a bucket with a showerhead. Unscrew for water and tighten to stop. Simple."

"That does seem straightforward," I said with some relief. *Familiar but with a twist, literally.*

We returned to the kitchen. "Let's put your stuff in your house, and then we'll have a welcome drink in the living room," Karen suggested.

"Sounds good." I picked up my backpack and tetra-sack, then followed Karen to a small bamboo hut thirty feet from the kitchen.

"This took one day to build, and fifteen dollars that you can give to me anytime," John said as he opened the door and placed my backpack on the twin-sized sleeping frame. I looked around and saw some nails that would function as hooks to hold my clothes. The floor was the same dead coral that was everywhere in the compound. I left my bags and followed John and Karen into the main house, adjacent to the kitchen. The living room had a split bamboo floor, bamboo benches against the walls, and a large trunk that served as a coffee table.

"Karen practices her songs here, and it's where we entertain visitors," John said. "In the back are the bedrooms." Their beds were like mine, up on bamboo platforms supported by sturdy branches nailed together and topped with thick, spongy mattresses. There was a second bedroom with a twin-sized bed for their four-year-old son, David, who had fully assimilated into his peer group. He often stayed at Jole's house playing with Alan and Donald, Jole's younger sons. David rarely went to Santo with John, preferring to stay on Malo with his friends.

John offered me a beer and I happily accepted it. We made small talk until I could no longer suppress a yawn. On my first night, I went to sleep utterly exhausted but relieved to have many of the comforts of home. I was surprised by how much cooler and more comfortable the elevated, slatted bamboo frame was than the Western-style beds in Santo.

The next morning, I put on my Abercrombie outfit and prepared for the day ahead. I was the first one up, so I went for a walk. I followed a path with no idea where it led. There were stands of tall coconut trees near every house. Around those trees were fallen coconuts. I had been told about the importance of copra, dried coconut meat—the primary source of cash on the island.

To make copra, the white fruit is scooped from the shell and placed on a platform over a fire. The heat from the fire dries the coconut, making it ready for export.

As I walked, I could see that one of the copra dryers had caught fire. As I got closer, I approached a group of men sitting nearby. I didn't know what words I could use to alert them to the fire, so I pointed and shouted, "Fire, fire!"

They looked up and smiled at me. One of them came over and explained that it was fine—or at least that was what I thought he said. I watched a bit longer as they quickly returned to their conversation. I would learn that very little could interrupt Malo men when they were socializing. Apparently, the fire in the copra dryer was nothing to worry about. *Another day of living and learning*, I thought as I retraced my steps back to my hut.

FIGURE 9: Avunatari Mission House

FIGURE 10: Copra dryer. The fire burns under the platform, and the empty shells are set on the side.

When I returned, I decided to try out the shower. I had to fill a bucket with water from the tank, heat a kettle of water on the propane stove, add the boiling water to the bucket, carry it to the shower, and hang it in the shower stall. Then I could undress while the water cooled. This was all straightforward and functional. I turned on the showerhead, and the warm water felt so good and familiar that I forgot where I was. I lathered up and then ran out of water. Embarrassed, I called for help. Fortunately, Karen came to the rescue, but it took ten minutes of watching the soap cake on my skin before the refilled bucket arrived. My first shower took an hour and four buckets to get clean.

After that incident, I remembered to turn the showerhead on, get wet, turn it off, wash myself with soap, turn it on again, and rinse off the soap. In time, I became efficient. I had three gallons of precious warm bucket water and made it work down to the last drop. I bathed using less water, but I enjoyed the full experience. It was as close as I got to the warm, steam-filled showers I loved back home.

Washing my hair was a separate process, which required a basin, and was best done in the heat of the day. The process was simple: Fill the basin with water (unheated); dunk hair and head into the basin; remove and shampoo; rinse by pouring half of the basin water down the back of my head; arch backward to flush the remaining water down the front. Conditioner and other amenities were not practical. The process was also tedious, so shampooing once a week became my new norm.

The housing situation was pleasant and cleverly set up. Despite the doctor's fears, I could and did drink the water. Every night at dusk, I'd hear *sputt-sputt*. It was John starting the generator. When it worked, we had the only electric lights on the island: Bare, incandescent hundred-watt bulbs hung in the kitchen and around the living room. They flickered whenever the gas-powered generator sputtered. When it quit, which was frequently, John would light the kerosene lanterns. In the evenings, we'd gather around the old trunk in the intimate living room. Sitting on the bamboo benches, we'd chat, sing folk songs, or play cards with the generator chugging in the background. It was not uncommon to begin the evening in the glare of hanging white light and finish in the soft yellow glow of a handheld lantern. It was elegant camping.

John had been a submariner in the navy, where he learned basic organizational and electrical skills, which helped explain why he had a generator. He had another clever device, which he called a "battery box." In the wooden box were twelve D-cell batteries and a cable that connected to and powered the radio. The batteries lasted for months; it was both efficient and cost-effective. At the time, I was focused on what was different from Philadelphia and didn't fully appreciate the high-end setup he'd created.

Karen liked to play card games, and she loved to win. I also love to win and have never let someone win because it was expedient. If she lost too frequently, the offending card would be torn in two. The cards were victimized out of her frustration at losing, and possibly from being cooped up on Malo. Fortunately, her fits of pique were short-lived. We created new cards, first from the jokers and then from a previously compromised deck. By the time I left, more than one-third of the cards in our deck were substitutes.

Every morning, I'd wake to the sounds of the mynah birds "talking" in the coconut trees or the clatter of pans from Jole's house. I'd dress, usually in jeans, pour water over ground coffee, and let it steep while I scrounged for food. After five minutes, I'd strain the coffee and enjoy the hearty hit of caffeine. At the kitchen picnic table, John would talk about his plans for the day. By eight o'clock, we would be on our way. Karen would still be asleep, and David would head next door to play with Jole's kids.

I followed John. We'd always run into people on the road and spend time talking with them. We'd frequently pass gravesites made from large pieces of coral set upright in an oval surrounding a flat area where the body would be buried. I'd make notes and transcribe Tamambo words phonetically. Vira, John's foreman and a Malo native, provided the terms and translations. Occasionally, we went to an ancient Lapita site, but there was little to see. A reddish piece of pottery was an exciting find.

Between a Rockfish and a Soft Landing

After a few weeks, I asked Vira to show me a stonefish. He chose a windless night with a full moon. Wearing tennis shoes, I walked with him out onto the reef. He stopped when we came to a water-filled passage in the reef.

"Look," he said softly. "There is a stonefish." He pointed to some coral with the tip of his spear.

"Where? I don't see it."

Vira took a flashlight out of his pocket and pointed the light at the spot where his spear tip was poised.

"There," he said. "Right here." Then he plunged his spear into the water and brought up the dying stonefish. It was a mottled brown color and encrusted with marine crud. We walked back to shore, and he put it down on the beach to die. I knew I would never be able to see or find one on my own, which reinforced the practice I had adopted of never touching the reef while I swam.

John encouraged me to explore the environment with Vira. John was supportive and sociable, and he introduced me to people, giving me a soft landing into island life. Still, I was overwhelmed with new people and new ways. It was a fascinating world of subsistence agriculture—no stores, no fast food, no television, no books—just people and families. Each night, as I lay in bed, I thought about what I'd learned and reminded myself, *At least I didn't die today. At least I didn't die!*

Lessons Learned

- Schooling gives status, even on small remote islands.
- You don't need fancy Western plumbing to have a nice shower.
- Hair survives with only weekly washing.
- Living is much easier when you take advantage of someone else's experience.
- A soft landing in a foreign place makes it easier to adjust.

Chapter Three:

A Dilemma Dissipates

There were no sea snakes as far as I could see (which was about one foot without glasses). I was surviving and acclimating.

John gave me my first Bislama language lesson: "Here are a couple of good phrases for you to know. The first is *me no savvy,*[1] or 'I don't know.'"

"*Me no savvy*?" I repeated. "So, there is no word for 'I'?"

"Correct. About 95 percent of the words in Bislama are English, with a smattering of French and Portuguese. When in doubt, just say '*me no savvy*.' Another helpful phrase is '*thank you too mas*,' and if you want something, just say '*me wantem*' followed by the English word. They'll understand."

"Got it." I wrote these simple phrases down in my notebook.

"One more point," John added. "The grammar is very simple and uses the structure of their native Tamambo language. So, everything you say is in the present. You indicate past and future by adding the words 'finished' or 'tomorrow' or 'sometime.'"

"How does that work?" I asked.

"'I went to town' is *me go long town finish*. 'I will go to town next week' is *me go long town week em he come*. It is reasonably straightforward."

Even though it seemed easy, it still took a couple of weeks to tune my ear to the local accent and parse the sentences. It took another week or two to be comfortable speaking simple sentences. After three months of trial and error, I was fluent. Step by step, I acclimatized to life on Malo.

---

1. The phrase is written phonetically as *mi no save*. However, for ease of reading, I use the English spelling for common phrases such as "thank you too much" (*tangkyu tumas*). The name of the language, Bislama, comes from the French *bêche-de-mer*, which describes the sea cucumber that is found here and exported as a delicacy.

The island wasn't on anybody's radar until World War II, when Espiritu Santo became the staging base for the Battle of the Coral Sea and the largest advance base in the Pacific, frequented by Admirals Halsey and Nimitz. According to the recently opened World War II Museum in Santo, Charles Lindbergh arrived in 1944 as a private citizen to give flying instruction to novice pilots stationed there.[2] Although Malo was peripheral to this activity, some islanders went to work for the US Navy.

When locals learned I was American, they talked fondly about the Americans they'd met during the war. John's navy experience gave him a common bond with the men who had experienced the American presence. The Malo Islanders would volunteer information about grave sites, geographic anomalies, and legends.

Piecing It Together

John's archaeological research involved digging trenches and finding pottery pieces.

"Here's a piece of Lapita pottery," John said one afternoon.

I took the reddish-brown fragment.

"It's between two and three thousand years old," he said.

That's cool, I thought as I examined the shard resting in the palm of my hand. It was smooth on one side and stippled with lines on the other. "Isn't Lapita associated with the ancestors of the Polynesians and Micronesians? People who migrated through New Hebrides from East Asia?"

"That's what we think. New Hebrides was a crossroads. Culturally, the people share many similarities with Polynesia, including the language, food, cooking practices, the way houses are built . . . lots of overlap. But physically, the people today look more like the Indigenous people of Papua New Guinea (PNG) with their darker skin, and there are artistic and cultural similarities with them as well. We assume the Lapita people were the first settlers, because there's an archaeological record of their presence. This is what makes it so exciting for me. I am piecing together the lives of the first inhabitants of Malo."

Literally, piecing together from pottery chips, pieces of charcoal, and patterns of stone, I thought. I admired John, his persistence and ingenuity. His enthusiasm helped make the days pass quickly. Nonetheless, as an observer, I was bored collecting these scraps from ancient lives.

---

2. Malo World War II Museum exhibit.

Figure 11: Lapita pottery. Pot reconstructed from potsherds. On display at the Vanuatu Cultural Centre. Used with permission.

"How interesting," I would say every time he showed me a stone circle, but it really wasn't. He was keen, but I was not. I had learned from my brief experience at a Hopewell Native American site near Cincinnati that I would never be an archaeologist. Weeks of sifting dirt and cataloging fire-cracked rock were beyond boring. Real-life archaeology was rarely the drama-filled, golden-treasured excavations of the movies.

Okay, I thought, *beaches are good. Nobody's needed the tetracycline and I don't think they ever will, even though it's now at the dispensary. I need to think of something I can do that is interesting.* I had to find a dissertation topic—one that I could leverage into an academic career.

As we walked around the island looking at Lapita sites and burial plots, we would talk with the islanders. Everything was new and different for me: the people, vegetation, climate, sounds, smells, foods. There was so much to absorb—it was overwhelming. *You'll need to find a way to describe this if you're going to be an anthropologist*, I told myself. When we met locals along the road or near a dig site, they were happy to talk about their traditions and history. I was delighted to listen.

The ni-Tamambo would tell stories about their local gods, the ones in the coconut trees, the ones up on the plateau or in the ocean, and they could describe the origins of these gods. They knew whether islanders had come from other nearby places, like South Santo, or had been on Malo "forever." Wanting to connect with and learn about people, the first questions they asked me were invariably about life circumstances and family: "Where are you from? Are you married? Do you have brothers and sisters? How many? Are your mother and father alive?"

"I'm from California. I am not married. I have three brothers, and my mother and father are alive," I would reply.

"Do you know Peter from California?" It was common for me to get a question about someone they had once met, often during the war.

"No, sorry, I don't know Peter. Do you know what city he was from?" That usually stopped a line of questioning, but not their search for a way to connect, either through common acquaintances or family structure. In the course of a conversation, one would learn the other's genealogy going back for generations.

John liked rocks, stones, and pottery shards. He got excited by them and the stories they told, but Malo was a place for cultural anthropology—I could see that, and it was why I had come.

Can't Get No Statusfaction

In the 1970s, the language between expats and locals was jarring. In Santo, I felt I was walking through a world unchanged from the nineteenth century—an anachronistic refuge from the blackbirding days when Melanesians were lured and kidnapped to work in the cotton fields and sugar plantations of Queensland. In 1973, a conversation between an Indigenous man and an expat would go something like this:

"Massah, you wantem me pick-em up boxes belong you?"

"Boy, em he good. Thank you too mas."

Every time I heard the "massah-boy" exchange—a dichotomy that began in Australia, along with Bislama—I would cringe in disbelief. It was an atavistic colonial stereotype I never expected. These exchanges were uncomfortable. When I returned two years later, the offensive, subservient language had evaporated. Everyone—expats, locals, and natives—referred to each other by name.

The reciprocal terms "massah" and "boy" reflected status, which was important both in Santo and on Malo. Whenever we went into town, where there was a mixture of expats (mostly Australians and French), the structure felt more European—in fact, my status in Santo was low, since I was a single woman. Nobody cared about my academic credentials. On Malo, I had the benefit of John's professional status. In Santo, there was no halo effect.

I became more aware of this colonial way of thinking as time went on. The color of your skin, your links with other places, what you had access to—they all determined how you were perceived. Of course, for me, the fact that I was a woman and American mattered most. I was there, in their eyes, because of John. I was there to help him, and I was an oddity—single, alone, not defined by any relationship to a man other than him.

On Malo, my status as a white woman was higher than that of a native woman, but a woman always had a lower status than a man. At events, I would initially go to the women's side but could easily walk over to the men's side and converse. However, I always followed the gender-appropriate protocol with those men who adhered to the island's traditional customs. I was careful, mindful, and respectful in their company, especially when I was told it was taboo for me to eat food cooked on their hearth or even touch the stones of that hearth. The consequence of violating the taboo could be severe and even fatal. So, I restrained my urge to prove a point as it did not seem worth discovering the consequence of breaking this *kastom* law.[3]

By Malo standards, I was a rich, educated white woman—and there was a fine line between what the richness, the whiteness, and the education brought as opposed to my gender and ignorance. What status I gained from the first three factors was never enough to negate being a woman, but it was enough to let me move freely between groups of men and women. This freedom gave me a sense of confidence that also allowed me to be vulnerable when I tried to learn the local customs and language. Initially, conversations were superficial, limited by my linguistic ability and ignorance of what questions to ask. Culturally, I was a toddler trapped in the mindset of an accomplished American adult.

I reinforced my individuality by wearing jeans. Jeans were a good compromise as they were masculine in a gender-conscious environment and they differentiated me from a stereotype. Native women wore something similar to a muumuu, an island dress called a Mother Hubbard.[4] These dresses were introduced into New Hebrides in the 1800s by Presbyterian missionaries to

3. *Kastom* is a Bislama term meaning the traditional (customary) ways things were done: the traditional culture, economy, agriculture, etc. It is synonymous with the traditional, precolonial way of living. The term is ubiquitous throughout Melanesia.

4. These garments were named after the style of children's smock worn in the nursery rhyme of the same name.

cover up the local women's perceived nakedness. Women sewed their own dresses from large pieces of colorful and often floral-patterned fabric. (I wasn't tempted to wear one—it seemed wrong and premature. Jeans also made me feel more comfortable.)

On Malo, children were naked for about three years, then clothed by the time they went to school. Girls would wear home-sewn shifts as this was a culture in transition. They had moved on from the tradition of wearing only a pandanus/hibiscus fiber skirt or pubic coverings made from leaves. It was accepted that women should wear more modest coverings than those of their ancestors, or even their parents in some cases. Most men, including John, wore Chinese shorts; fewer than twenty of the islanders (0.1 percent) wore the traditional woven loincloths. John, however, was one of the few men on Malo who routinely wore a shirt. Island clothing was practical, inexpensive, and easy to wash. The latter was especially important, since the chores of daily living relied on a lot of physical activity.

Tasting Man

The ni-Tamambo practiced subsistence slash-and-burn agriculture with traditional tools and had an intergenerational transfer of their extensive local knowledge, customs, flora, and fauna via oral tradition. Culture change happened slowly. Their subsistence-based economy allowed them to survive without money. It felt similar to what Captain Cook would have encountered in 1774 when he named the island St. Bartholomew.[5] Even when they had modern conveniences, they retained their traditional survival skills—everyone was familiar with matches, but they could easily start their cooking fire using a wooden hand drill and coconut fiber. Many people had battery-powered shortwave radios. I absorbed it all.

"There are *kastom* people on this island," John told me. "Pig killing is still an important part of their ceremonies, and this is something you'll hear and learn about."

And one more thing . . . cannibalism.

It turned out the occasional "leg of man" was something that ran through the history of Malo. It remained a topic of conversation for many people,

5. Elizabeth Reed Dickie, "A Note on the Naming of Malo Island, Vanuatu," *Journal of the Polynesian Society* 90, no. 1 (1981): 111–16. The island was recognized as Malo when Commodore Goodenough stopped on the island in 1876 and asked the natives what they called the island. "Malo" was the response he got from a native who spoke English.

especially the elders. One day, when walking through Avunatari Village, John made a detour to a traditionally styled hut.

"This is Anji's house. Anji is the oldest person on Malo, and he's reputed to be more than one hundred years old," John said.

"How do they know?" I asked.

"He says he remembers being a boy when the first missionary came. He also says that Antas told him stories about his trip to Australia with Reverend Landels. It's at least consistent with being one hundred. Everyone else who's still alive says that Anji is older than they are."

"That's impressive. So at a minimum, he's at least ninety, since the missionaries arrived eighty-six years ago."

"Yes. Let's talk with him," John said. "He doesn't leave his house much anymore. You'll see."

After receiving an acknowledgment from inside Anji's hut, we entered and sat on a woven pandanus mat. Next to us was a wizened, frail old man with kinky hair and a long white beard, closely cropped. His face was framed by bushy white eyebrows and a matching mustache. He wore a white long-sleeved shirt and loose trousers.

After a brief introduction, John bluntly asked, "Anji, did you eat people?"

"Did we eat people?" Anji repeated back. "Yeah—and they taste gooooood!"

I got the impression that Anji reveled in his celebrity as the oldest person on the island and may have exaggerated how old he was. In any case, no one was bothered by those details. Anji would happily talk about cannibalism in pidgin English all day long.

Was he playing to the gallery or was it genuine? Who knows, but it was interesting and consistent with what I learned later—that there were cannibals on the nearby island of Malekula, and on other islands, missionaries had occasionally been consumed after meeting their demise. It occurred frequently enough that New Hebrides was called the Cannibal Islands. The same moniker was also applied to other islands in Melanesia.

Malekula is the archipelago's second-largest island. Shaped like a sitting dog, it is just south of Malo and was, from what I'd heard, stuck in time. The inhabitants still performed the rites and traditions of their ancestors. They also had "kitchen sites" where they prepared cannibal meals. When I later checked, expecting all of this to be a practice shrouded in the mists of time, I discovered that the last recorded act of cannibalism on Malekula had been in 1969—just four years earlier. Although that island received some notoriety as the center of human flesh–eating, it was practiced on many of the other islands too.

Despite my concerns about all the ways I could die on Malo, I hadn't thought I might be the main course—it just hadn't occurred to me that there would be cannibals. As Anji told his stories, he would elaborate on the old days when, as part of the *Sumbuea*, high-ranking men would eat Malo's enemies.[6]

Anji would become more and more engaged as he told his stories. Wide-eyed with incredulity, I sat with John and listened to his tales.

Figure 12: *Kastom* (custom) men at a wedding. Bani Vahalima (front left) was the most traditional and highest-ranked *kastom* man, and Muele (far right) was a *kastom* man who kept many of the traditional dances and stories alive by sharing them with the non-*kastom* natives.

6. According to anthropologist Robert Rubinstein, the *Sumbuea* "was a gradual, orderly, sequenced device for the expression of male selves . . . It was . . . a system of publicly counting coup on any number of items which a man had brought under his controls. It appears that in the past, human sacrifices were the ultimate item in such a system. Moreover, it seems from the few stories I collected about cannibalism that humans might have been the only item of 'meat' that could be shared by men and women of a number of different eating classes (grades) . . . it also appears that such occurrences were rare . . . or did not provide a central focus. . . . The *Sumbuea* developed around the pig." From Robert L. Rubinstein, "Placing the Self on Malo: An Account of the Culture of Malo Island, New Hebrides" (PhD diss., Bryn Mawr College, 1979), 259.

FIGURE 13: Men's *nakamal*, or sacred hut

Nuptials and Circle-Tuskers

In addition to Anji, John and I visited other elders. The islanders who lived near John's house would come to his home. He always had liquor, so everyone would have a beer as they chatted.

Every week, John would drive his boat to Santo, occasionally accompanied by Karen, who sometimes sang at a local bar. John would replenish our supplies and bring mail back to Malo. He also socialized—he had close ties with some of the expats in Santo. To islanders and expats alike, John was a "rich" person who had the freedom to come and go as he pleased. In addition, the islanders considered him a *big man*, the Bislama term for someone of influence and with largesse to distribute.

I observed and learned from John, often taking notes. I also paid attention to local customs. At weddings and other important events, men and women arrived together, often with the women and children leading the way. Once at the venue, men and women always separated to opposite sides to be with other members of the same gender. Their roles were clearly defined.

At a wedding, there would be a gift exchange between the bride's and groom's extended families. Pigs, yams, and trinkets moved between people in a choreographed way, presumably reflecting lineage, customs, status, and hierarchy. The transactions were visible and obvious, yet the underlying rules seemed complex and obscure.

Jole and Vira did their best to explain a typical wedding. It was a mixture of both Christian and traditional elements. The balance depended on whether the local pastor was officially involved in the ceremony. It was messier than the clear patterns described in ethnographies. For me, an ethnography of the ni-Tamambo was not going to be easy.

I used a later wedding as an opportunity to be an ethnographer—a cultural anthropologist. My goal was to follow the journey of a single yam brought to the wedding. I chose a distinctive-looking yam with a nose shaped like a manatee. I tracked "my" yam when it was placed on the pile of yams. I watched as it was picked up by someone I thought was a cousin of the bride. Then, I lost it. I never knew if it went into *laplap* or was given to another family member or was planted. I questioned whether I was cut out to be a cultural anthropologist.

I asked myself, *Even if I had followed that yam, what significance did it have? What are the rules?* I was certain I was witnessing what I'd studied in my classes—gift giving, exchange, and redistribution with the nuance of who gets what and of what size and/or variety—but I could not for the life of me see why it would be of the slightest interest to anyone. It didn't even seem to interest the people who did it.

In contrast to the yam, pigs were of great interest to the attendees. I watched the groom's family escort a pig to the front of the house where the bride's family waited, men on one side and women on the other. The pig was presented and then slowly walked off to the periphery, where it was ceremonially killed with a hand-carved wooden club. There was much more to the ceremony than a simple slaughter for food. The pigs themselves were special. A parade of circle-tusk pigs signified an important event.

Malo was famous for breeding circle-tusk pigs, but why were they sacrificed at important events, like weddings or taking a grade and moving up to the next level in the *Sumbuea*? A circle-tusk pig was always male, and its lower tusk grew throughout its life. Early on, its upper canine teeth were removed so the lower tooth's natural curve was accentuated when it reached the upper jaw. Continuous pressure on the tusk bent it and it penetrated the lower jaw, forming a complete circle. It took seven years for one circle to grow, making the pig an object of wealth; the circle showed that the owner had other resources and didn't need to slaughter the pig for food. If the pig's tooth made two circles, it was an even greater sign of wealth.

FIGURE 14: Circle-tusk pig

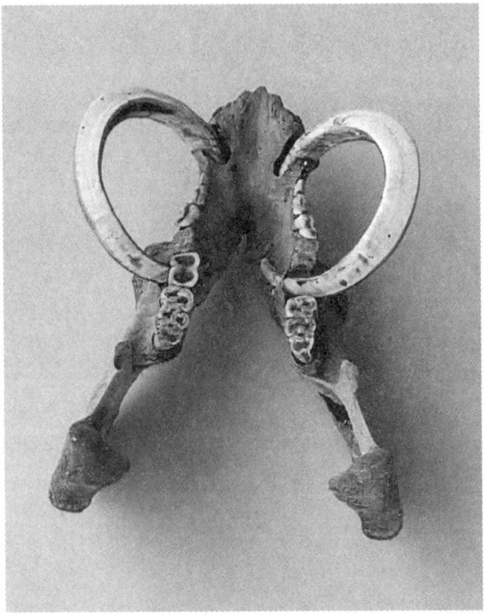

Figure 15: Circle-tusk pig jaw. Citation and permission received from Matthew Spriggs ("one of the many pigs killed by Chief Richard Leona of Loltong Village, North Pentecost").

Pigs roamed freely on the island, as did chickens, cats, and dogs. Only those destined to be circle-tuskers were tied to trees and fed by hand. These pigs ate only mature coconut seed, which looked like a snowball when the coconut was opened.

Malo was part of an interisland trading ring for circle-tusk pigs, but the hermaphrodite pig was unique to Malo.[7] A circle-tusk *narawe*, the most valuable of pigs, had both sow genitalia and low levels of testosterone.[8] At the time of my stay, about 20 percent of pigs on Malo were *narawe*. Malo is generally accepted as the original source of these pigs, which were traded to the adjacent islands of Espiritu Santo, Aoba, and Malekula. Sacrificing a *narawe* was critical for men to advance in the *Sumbuea*.

Whose Dream Is This?

I was living the cultural anthropologist's dream. Everything I'd read about and studied was happening right in front of me. It was mine for the describing. I had my notebook and I wrote down what I saw. I talked to people and asked what was happening. I had crude diagrams of kinship charts with notes under the eldest uncle and father of the bride. I drew arrows between family members with symbols for money, yams, and other gifts. I tried to distill it into a dissertation topic. It wasn't coming easily.

I tried to write out my observations on daily life, but they were often trite and banal. It was hopeless; it just wasn't me. After two years of preparing myself to be a cultural anthropologist, I discovered I didn't have the patience or inclination to delve into the multidimensional nuances. I was happy to participate, experience, live in the moment, and flood my senses with Malo. Capturing this in a dissertation—transporting readers into this culture with an understanding of who these islanders were and how they lived—was beyond me. A travelogue, maybe, but an ethnography? No. I had read ethnographies that said "the X tribe does Y," but I needed to understand why they did that along with the antecedents and consequences.

---

7. The interisland trading ring was primarily between Malo, Aoba, Malekula, and Vao, a satellite island of Malekula.

8. James K. McIntyre, "Investigations into the Relative Abundance of Intersexual Pigs in the Republic of Vanuatu," *Science in New Guinea* 22, no. 3 (1997). Reprinted in James K. McIntyre, "Intersexual Pigs of Vanuatu," *The Southwest Pacific Research Project*, 2016, https://www.swprp.org/intersexual-pigs.

I needed details to make it come alive for me, and as an outsider, it was far too difficult to achieve this level of understanding. Within a culture, people act with behaviors that are contextually appropriate, but as with language, they aren't necessarily aware of the rules. I rationalized that if I *could* get to the point of contextual understanding, I would be fully assimilated and would have therefore lost my objectivity. It just wouldn't work. And if I couldn't get deep enough to truly understand, then maybe it was time to do something different.

I kept asking myself, *How in the world can I write this up? How can I possibly describe their culture in a way that is meaningful? How can I bring it to life?*

It was becoming clearer and clearer that I'd never had a real passion for cultural anthropology. I'd applied to Penn because of its paleoanthropology resources so I could study human fossils. I loved the concept of paleontology, though I realized it could be limiting. I really liked my adviser, Dr. Goodenough, and my fantasies of island life, so I had followed his interests instead and found myself on Malo.

FIGURE 16: Laplap: Step 1. Grated yam with added coconut cream.

FIGURE 17: Laplap: Step 2. Laplap wrapped in banana-like leaves and tied with spines from leaves.

FIGURE 18: Laplap: Step 3. A woman uses bamboo tongs to place stones on laplap, which rests on a bed of hot basalt rocks.

FIGURE 19: Laplap: Step 4. Completely covered with stones—baking.

FIGURE 20: Laplap: Step 5. The laplap is cut and served on the leaves, with or without fish, crab, pork, or rice.

I had gone to Malo certain that I would have my first taste of fieldwork and return to Penn ready to write up my dissertation proposal. I was sure I'd be primed to return to the field so I could research and write an ethnography on the Malo Islanders—even though this was exactly the type of book I avoided reading.

The brutal truth was that I didn't have a clue as to how I would describe this culture or any other. I could see people behaving differently from Americans, and the islanders explained to me why they did certain things. But structure? That was hard to see in real life.

Never mind that my interest in cultural anthropology had been focused on societal systems and social structures. I loved playing with Claude Lévi-Strauss's analytics to the point that I studied mathematical logic so I could better understand the basis of his paradigm, based in part on the Hegelian dialectic of thesis-antithesis-synthesis. I enjoyed figuring out the "rules" to understand Lévi-Strauss's structural framework for kinship.[9] I even took a one-on-one seminar with Goodenough to learn componential analysis.[10]

What I'd learned was that the derived structures I had once embraced as "fact" had no intrinsic basis in reality; rather, they reflected one's talent to create arbitrary rules that, when applied, aligned reasonably well with one's observations. It was circular logic—an untestable tautology. An example of arbitrary classifications is race; there can be between four and seven races depending upon who's doing the classifying. Additionally, race can include physical as well as behavioral traits—also depending on who's doing the classifying.

Where did this leave me? I tried to decipher their social structure. I cobbled together what I thought were the salient takeaways from reading Émile Durkheim and Max Weber, two of the founders of modern social science: i) It takes more than one person to create a culture; and ii) You know culture when you see it. Clear. And I was not seeing it. Thus, I was not going to be the person to describe Malo Island culture, or any other culture.

---

9. The Hegelian framework stresses that there is a proposition (thesis) and its opposite (antithesis). Synthesis is the reconciliation of these two extremes. Lévi-Strauss used this approach to analyze myths and develop rules for marriage in different cultures. See Claude Lévi-Strauss, *The Elementary Structures of Kinship* (Beacon Press, 1969).

10. Both Goodenough and Lévi-Strauss delved into abstract concepts to describe cultural "rules." Neither is cited much in contemporary anthropology. Lévi-Strauss believed there was an underlying structure in kinship systems and myths that was universal.

So what *was* I going to do? Cultural anthropology was not a good fit for me, and archaeology was a nonstarter. I had also studied and analyzed enough linguistics to last a lifetime (or so I thought). That left only physical anthropology. I was back to where I'd started when I applied to Penn: fossil humans in Africa, or maybe fossil primates in Wyoming. Well, at least there were beaches somewhere in Africa.

Most importantly, what would I do with my remaining time on Malo? I'd worry about what would happen when I returned to Philadelphia later. One step at a time—and besides, there were still more cannibals to meat (oops, meet).

Lessons Learned

- Listen to yourself and follow things that interest you. It is important to recognize what you like and what you don't like.
- Skin color, because it is obvious, can be used to "color" perception.
- Colonial thinking still exists.
- It is odd to be a single woman with no clearly defined identity.
- When in doubt, dressing out of the norm makes it harder to be stereotyped.
- Analytic paradigms, e.g., Lévi-Strauss's structuralism or Goodenough's componential analysis, have no basis in the real world.
- I was not going to be the person to describe any culture.

Chapter Four:

Island- and People-Hopping

I visited the island of Malekula three times that summer. Just south of Malo, it was traditionally an important trading partner with a reputation for having many tribes that were still *kastom*. My first visit was with John and Karen. We packed ourselves into the same eight-seater Islander that I'd originally arrived on for the forty-minute flight from Santo along the Malo coast. We flew above the palm trees to Norsup, a grass landing strip at the island's northern end.

That evening, we were going to a party given by Barry and Lesley Weightman in nearby Lakatoro. Barry was the agricultural officer for the Central District, and the couple were known for their parties. John's connections got us a free stay at the British Guest House, where I met George, a VSO attached to the agriculture department.[1] He was British, slim and cute with curly brown hair. There was an immediate attraction. At some point, our conversation turned to motorcycles.

"I don't know how to drive one, but I've always wanted to learn. Can you teach me?" I asked.

"Sure—it's easy. My bike's over there. It's very basic."

Together, we went over to the small motorcycle.

"Hop on and I'll show you the controls," he said. "Clutch, gas, brake . . . they're all here on the handlebars."

He led me through a dry run, and it seemed straightforward. "Let's go for a ride," he said as he hopped on behind me.

1. The Volunteer Service Organization, the UK equivalent of the Peace Corps.

It was difficult, but I managed to do some small rides with him on the back seat, ready to grab, clutch, gas, or brake whenever I did something wrong.

Later that afternoon, I needed to go back into town before the party.

"Here, why don't you take the motorcycle?" George suggested.

"I don't think that's a good idea. I'm really uncomfortable driving it. Is there a car I can borrow?"

"Sure, take the department's Land Rover." He went into a nearby building and returned with the key. "Here you go. You know the way?"

"I think so. Just back up this road and follow the occasional signs to Lakatoro."

"Right. One other thing. Do you know how to drive a manual transmission?"

"Yes. My car at home has a stick shift."

Keys in hand, motor started, and transmission engaged, off I went, driving quite confidently, and maybe a tad aggressively, on the single-lane coral-gravel slalom course through the coconut trees. All was fine until I could no longer steer and the brakes stopped working. The truck skidded lazily until its forward progress was impeded by the trunk of a palm tree and came to a complete stop with the engine still running. I turned the key off and the engine still kept running. I put the transmission in neutral and it kept running. Desperate, knowing there had to be a way to turn off the truck, I pushed every button and pulled every knob. Finally, the diesel gasped to a stop. I got out and looked. The coconut palm was fine, and the truck didn't look so bad, either.

I got back in and tried to start it. I put the gear shift in neutral. I turned the key. Nothing. I tried the smaller transmission but, again, it didn't start. In frustration, I counterintuitively pulled a knob labeled "Choke" and turned the key. The engine came to life. Fortunately, the rest of the slow drive was uneventful, allowing my adrenaline to return to normal.

When I arrived in Lakatoro, everyone was very kind and said the damage was minor, so the truck could still be driven. The replacement parts would probably arrive from England in a couple of months, but not to worry.

What a party it was. Barry and Lesley were costumed as King Arthur and Guinevere. Through a loudspeaker, their sound system blared songs that were popular for dancing. George and I danced for the better part of the evening and ended up singing along to Neil Diamond's "Sweet Caroline" at the top of our lungs. It was a great party and a fun weekend.

June 25, 1973
Dear Mom,[2]

Tomorrow we are leaving for Malekula to buy a collection of Small Nambas artifacts. John and I plan to have an exhibition in the museum when we get back, and all of this stuff is very much in demand. The museum has offered to buy it, but we are having to use our own money to procure it, thus the extreme lack of money on my part.[3]

I have gotten to know myself better thanks to this experience, and Goodenough is now chair of the committee of a physical anthropologist in the Pacific. There is no way that I really enjoy doing cultural anthropology, and I came to grips with the fact that why be a mediocre cultural anthropologist when I could be a first-rate physical anthropologist? My interest is in the more objective biological and physiological characteristics of people, something I think Goodenough has known all along, and I think that was one of his main reasons for sending me here now, rather than for a year with my degree hinging on it. I also think they gave me the physical anthropology teaching assistant position knowing I would enjoy it more and do a better job at it. In fact, [Physical Anthropology Professor] Sol Katz said they would welcome me back to physical anytime, and there are not many physical anthropologists in the Pacific.

Thanks for writing so frequently,
Love, Betsy

Leg of Man

Just before I was to leave Malo, John and I returned to Malekula. He'd arranged for us to visit and stay in a remote village of what I thought would be the Small Nambas tribe. But it turned out that they were Big Nambas.

There are two different Nambas tribes—yes, the Big Nambas and the Small Nambas. In the 1970s, the world of the Big and Small (*Smol*) Nambas was unknown to outsiders. *Nambas* is the native term for "penis wrappers."

---

2. Unless otherwise noted, all letters were sent to the person named. I have made minor mechanical, grammatical, and capitalization changes to the letters transcribed in this memoir to improve clarity and readability. These changes do not alter the content or meaning.
3. John and I purchased artifacts that he shipped back to Penn. After that, I neither saw the sculptures again nor was reimbursed for my share of the purchase. Inquiries had no impact and no one had any idea what I was talking about. What happened to the art remains a mystery. (At least to me.)

One tribe wrapped the penis in woven and dyed pandanus matting that was then secured with a bark belt, and the other used a folded banana leaf that was also tucked upright into a belt. The *nambas* for the northern tribe were elaborate, whereas the banana leaves used by the group farther south were much simpler. One look and there was no doubt about tribal affiliation (see also Color Plate 4: Small Nambas hunting with bamboo bow and arrow).

FIGURE 21: Big Nambas man. Returning home after a reasonably successful wild pig hunt—Big Nambas area, interior of northwest Malakula (1973). Photo courtesy of K. Huffman.

FIGURE 22: Small Nambas woman. Small Nambas woman with traditional nose stick cooking bananas for breakfast—Small Nambas area, high interior of south-central Malakula (1973). Photo courtesy of K. Huffman.

The men wore their *nambas* when they went about their daily business, and it was considered perfectly normal and natural.[4] I didn't want to stare, but, of course, I did. After a few moments, the novelty wore off. Their dark testicles blended in with their dark skin. The effect was one of sameness, in contrast to the more familiar pink against white.

Traditionally, women didn't cover their top halves, and mothers would breastfeed whenever and wherever the baby demanded. Nobody cared—it wasn't a big deal. Women also had a little covering of leaves over their pubic area, secured with bark.

We stayed three days with the Big Nambas, who were among the last of the overtly cannibalistic peoples. They still talked about cannibalism in what seemed to be an excited way, and many of them acknowledged having done it. (On Malo, only Anji had confessed.) The thought that this practice had been widespread less than ten years earlier was both fascinating and unnerving.

4. For an example of how comfortable the men are, see the bonus feature "Going to Venice" in the movie *Tanna* (2015). It shows islanders in traditional dress very similar to that worn by the Small Nambas; these islanders casually walk and sightsee in the streets of Venice prior to the Venice Film Festival. *Tanna* won Best Director of Photography and the Audience Award for Best Feature Film.

It wasn't that islanders had simply been roaming around, snacking wherever (or on whomever) they liked. Rather, they practiced cannibalism when they killed enemies in warfare. Ritual cannibalism was part of the symbolism associated with victory. Most clashes were over land rights, and they believed they could absorb the power of deceased adversaries by eating them. Prisoners would be killed and their bodies butchered like any other mammal, then cooked in the *nakamal*, or men's sacred hut, like any other source of meat protein. The difference was that consuming an adversary conferred status. Around Malekula, even today, it is still possible to visit sites where men were butchered and see piles of long bones near the decapitation stone (literally "leg of man").

It was the men who talked about this. I never heard the women say a word about cannibalism. There was a hierarchy about who got what and which pieces of the body went where, and the Nambas men would discuss it. You could ask them what people tasted like. They always said the meat was good (or "yummy") and that it tasted like pig.

Anji had talked about cannibalism openly and with laughter, but there was more reticence with the Big Nambas. They were ambivalent about the practice; it improved their status, but they could be prosecuted for indulging, since cannibalism was illegal. Sometimes, I had to encourage them to tell their stories, because they knew it was no longer acceptable. I'm guessing the colonial government had made that change, and although its presence and influence were minor, there was still an understanding that cannibalism was not to happen—eating people was just not kosher. Islanders had been arrested and jailed for cannibalism; the penalty was real. In addition, the nascent Vanua'aku Pati political party (Our Home Party) had also put the kibosh on the practice.

Like the country itself, each district had British and French authorities who governed it—and either of those authorities could take action against a native who did not perform in a way the European powers deemed acceptable. The structure of authority on the islands was interesting. Even though I didn't see an overt government presence, I sensed that those in power would be alerted and might prosecute a person if a serious law had been broken.

This wasn't just in relation to cannibalism, but that was the topic that intrigued most people. Although it is such a taboo in American culture, it was actually something for which I had a reference point. In high school, I'd read *Ordeal by Hunger: The Story of the Donner Party*. The Donner Party was famous for resorting to cannibalism when a snowstorm trapped them in the High Sierras. It was a fascinating book, and the expedition's hero shared my mother's family name, Reed, which made me feel personally connected. Moreover, whenever we drove to Lake Tahoe, we went through Donner Pass and saw signs for Donner Lake. The tragedy of the Donner Party was part of our California history. I had been on my way to return the book to the library when Mom stopped me.

She looked at the book and said, "I've tasted human flesh."

"What?! How? In medical school? Did you snack on a cadaver in anatomy class?"

"No. A reporter offered me a dollar if I would eat a piece of charred flesh and tell him what it tasted like."

"Okay, Mom," I challenged. "How did the reporter get the barbequed flesh?" I wasn't sure whether to believe her, but I wanted to hear more.

"I was fifteen when the Cocoanut Grove nightclub caught fire and killed five hundred people. Our house on Concord Square was about a mile away. I was awakened by the sirens and when I went to the window, I saw the flames. I woke up my brother and our mother. Joe and I got dressed and walked toward the flames in the smoke-filled Boston night. There was burned and charred stuff scattered in the streets and the smell was awful."

"How close did you get?"

"Close enough to be scared. We were standing and watching the chaos when the reporter approached us and asked if we'd eat the scraps and tell him what they tasted like."

"Well, what *did* they taste like?"

"Like burned pig. You really couldn't taste much."

"Why did you decide to try it?"

"I thought, *Why not? I'll probably never have the chance again*. I was curious."

Her experience was unusual but didn't sound completely mad. To me, it seemed interesting—not a big taboo—and probably exciting given her age, the situation, and the era. It was also something I could imagine myself doing in the same situation. I've often wondered whether Mom's story played a role in my decision to become an anthropologist.

As an anthropologist, you are taught to accept things as they are; you are supposed to be unbiased and not think of how it "should" be. Cultural relativism is a cornerstone of the discipline. You try to understand things and events in their cultural context, through the lens of how that culture thinks they should be. Objectivity is key if you are researching another culture, and that was how I approached Mom's story.

Small-World Syndrome

As my time on Malo came to an end, I had more confidence in my ability to handle new situations and make friends with people from other countries and cultures. My time in New Hebrides had helped me realize how much I could do on my own by listening and observing and then incorporating the information. I learned to communicate in a new, albeit simple, language through trial and error. I also recognized that I had to find a new path to my dissertation. Naturally, there were moments of concern and anxiety, but it was all a learning process, and I was heading back to the States with more self-assuredness and resolve.

I island-hopped my way home to California. First, I flew from Vila to Fiji, and as I got off the airplane, an attractive middle-aged Australian asked if he could give me a hand with my backpack. Did I hear alarm bells? Nope, not one. It's hard to look back without the wisdom of the years, but I was young, I was out in a world where it was unusual to see single young women exploring, and I was reasonably naïve and always thought the best of people. Part of me imagined a worst-case scenario, but I also have an optimistic, adventurous approach to life—I wanted to see what would happen next.

"Where are you staying?" he asked me, and I told him.

"Oh, you don't want to stay there—stay where I'm staying!" he exclaimed. "It's next door and it's a nicer place."

Rather than running for the hills—or to my own hotel—I processed the information. "Okay, sure. Sounds good!"

Luckily, my story does not turn into one of abduction and terrible things. Peter turned out to be a world-class photographer who had my best interest at heart. In that same hotel, we met Mary Gehr Ray, an artist, and Bruce, an archivist who had been a student of the famous British anthropologist Alfred Radcliffe-Brown, the father of structural functionalism.[5] Among our newly formed group, I was the youngest by decades, so I went along with their suggestions. Over dinner, Peter and I discovered our itineraries overlapped

---

5. Structural functionalism was once very influential among British social anthropologists. The theory posited that social structures tend to evolve toward increased structure and increasingly specialize in the functions they perform.

in Tonga and Western Samoa. We would be together for the next five to six days. There were very few options for intra–Pacific Island travel, so it was not unusual for us to be on the same flights.

On Tonga, everyone chipped in for a rental car and the four of us drove around the island together. I was the ingenue with these accomplished people who were experienced travelers—it was a treat to be in their presence and to listen to their experiences. Peter taught me how to use a single-lens reflex camera. He explained the relationship between the aperture and the shutter speed and showed me how to get the right exposure by demonstrating the concept using my light meter. It was similar to learning how to use a slide rule: brilliantly simple and elegant once you understood the relationships. We would stop and take pictures of anything interesting. He taught me how to compose pictures while Mary sketched. She was small, unassuming, and talented.

Mary, Bruce, Peter, and I shared cocktails, talked about our favorite popular music, and explored the island of Tongatapu. It was a wonderful interlude. I took it all in and enjoyed the adventure. When I look back, I think how lucky I was—at that age—to have the flexibility to say "Okay, this looks interesting, let's see where this path leads." I realize I could have gotten into trouble, so much trouble, but I didn't. It was an amazing and memorable week of travel.

Next, Peter and I went to Upolu, the capital of Western Samoa. He rented a car and we explored the island. We stayed at the same hotel and always chatted with whomever we encountered. Occasionally, people asked us to dinner or suggested that we visit them. Wherever doors opened, we walked in. I took this for granted. I accepted and lived it. Samoans were nice, open, and welcoming, but it was also safe because Peter was with me.

Upolu is a typical volcanic island with a lush green shroud—another island paradise and one of Western Samoa's two larger islands. The beaches had sparkling white sand leading into turquoise lagoons. There were rainforests, rivers, and waterfalls. Every site we visited was postcard-perfect. Upolu was a great stopping-off point that showcased Polynesia's beauty. It was just that, though—a box to tick as I went from one idyllic place to another on my island-packed itinerary.

Next was American Samoa—this time, on my own. I followed Peter's teaching and talked to everyone, always being open to new experiences and jumping at opportunities to see or do new things. On one occasion, I was chatting in a grocery store while replenishing my peanut butter supply, and an American woman started talking to me. The next thing I knew, she invited me to a party that night.

"Why not come?" she said.

At the party, I encountered the small-world syndrome. Once she learned I was a Penn student, she asked, "Do you know Wilton Krogman?"

"Yes. His office is on the same floor as where I work. I've seen him in the hallway. He's famous for his work on human growth and development."[6]

"Did you know he was a twin and that the twin never developed and was lodged in his brain?" *No, I did not know that.*

I decided to play the game and brought up another famous Penn anthro professor. Loren Eisley[7] was a physical anthropologist and naturalist and kept the waxy yellow leg bones of a famous Philadelphia paleontologist in an umbrella stand in his office.[8]

We traded stories for the rest of the night. The last topic of conversation was, inevitably, *South Pacific.*

"You know, *South Pacific* was filmed here in Apia, and Western Samoa is the inspiration for the book, and Aggie Gray is the Bloody Mary prototype," she said.

"That's not what I've heard. The movie, sure. But Michener was stationed in Santo for three years, and the scuttlebutt is that Madam Ratard was his Bloody Mary."[9]

"Well, maybe. Regardless, you should at least go to Aggie Gray's for a drink. Maybe she'll be there and you can get her side of the story."

By the end of the evening, I had a next step and another door to open. I visited the rebuilt wooden hotel and I did have a drink. Unfortunately, there was no Aggie Gray to meet and no strangers intriguing enough to join.

The next stop on my itinerary was Tahiti. There I met a UC Berkeley zoology professor while I was wandering around Papeete. He was going

---

6. Wilton Krogman was one of the founders of physical and forensic anthropology. He was also a pioneer in primate dentition research.

7. Loren Eisley authored numerous books, including *The Immense Journey*, *Darwin's Century*, and *The Night Country*.

8. The bones were those of Edward Drinker Cope, who, in the 1880s, along with Henry Fairfield Osborn, advanced the tritubercular theory to describe the evolution of complex mammalian dentition from the simple, peg-like teeth found in reptiles.

9. It turns out that Michener's Bloody Mary was based on a Vietnamese woman who lived and worked at the Ratard plantation on Espiritu Santo, according to an article Michener published in the *Los Angeles Times* on October 19, 1986.

to Mo'orea the following day and suggested I go too. At the dock, I bought my ticket and arranged to stay for a couple of days at Cook's Bay in a small resort. I was the only guest. Most people spoke French, and I got by with monosyllabic answers courtesy of my high school French. Mostly I swam and hiked, enjoying the sights and marveling at the island's sharp green velvet-covered mountains. It was my time to decompress and relax.

I was exploring very well. I owe that red backpack a lot of thanks. And it was easy to spot going around the luggage belt. I'd bought it because I thought it would be easy to carry in the Melanesian "wilderness." On some level, I'd envisioned my Malo adventure as a camping trip, and the backpack resonated with that idea. Little did I know that it would be my ticket to so many new adventures.

I was safe the whole time. I took risks, but I didn't see them as risks then. It all seemed perfectly normal. I was meeting people who were working and living outside the box. This was 1973 in the South Pacific. No one was traveling there. There was no tourism and there were no cruise ships. It was relatively expensive to fly there, so you had to have a reason to go . . . unless you were painting or in the Colonial Service.[10]

The British maintained a presence in Samoa and Tonga, and there were missionaries. The Seventh-day Adventists and Mormons were proselytizing, so they had a reason to be there. But me? Not really. I had no vocational calling; it was a door that had been opened by paying an extra $24 on my round-trip airline ticket. After I had been to Fiji, Tonga, Western Samoa, American Samoa, and Tahiti, I flew back to America with precisely zero dollars.

The next step was to return to Philadelphia and see how the anthropology department would accommodate my epiphany about changing fields.

Lessons Learned

- Listen to others and learn to hear and see things for what they are.
- Seeing people unclothed is only interesting for a few minutes, and then it is natural.

10. In 1954, the organization's name was changed to Her Majesty's Overseas Civil Service. Colloquially, it was still called the Colonial Service.

- When you drive too fast in unfamiliar conditions, you crash.
- Only men like battle and cannibalism.
- There is accepted law and order wherever you go. It may be different from what you're used to, but it is there.
- If you are open and trusting, good things can come your way.
- A backpack is a signpost for adventure.
- Leftover money in your pocket earns no memories.

Chapter Five:

A Leap of Faith

There was no culture shock when I returned to the States, which was a relief. I decompressed at home in San Francisco for a few days before going back to Penn. On reflection, I knew that paleoanthropology would be risky. If I went down that route, my options were still the same: go to Wyoming, find a fossil primate, and spend the rest of my life describing those bones—boring. Or I could go to Africa as part of a team with a leader who got the glory while I would be digging the equivalent of Hopewell fire-cracked rock—also boring. I had no idea about what I wanted to do, only what I *did not want* to do. So I asked around and talked with my friends.

A senior grad student suggested, "Take a course with this professor." She pointed to a listing in the catalog. "He teaches only one course a year. He's good and the course is interesting. It's kind of in physical anthropology."

"This professor" was Dr. Baruch Samuel Blumberg. Born in 1925 in New York City, Blumberg earned his bachelor's degree at Union College while he was in the US Navy. After the service, he obtained his medical degree at Columbia, followed by his doctoral degree at Oxford University. Blumberg was an Associate Director at the Institute for Cancer Research (ICR), which is part of the Fox Chase Cancer Center. He led a large interdisciplinary research team that investigated differences in disease susceptibility and epidemiology. He also had an appointment in the anthropology department where he was teaching a course about the ways genetics, environment, and culture affect health. It sounded interesting, especially since I had enjoyed my undergraduate course in human genetics. So I thought, *Why not?*

Testing, Testing

I enrolled in Blumberg's biomedical anthropology course—and that changed my life yet again.

Blumberg was not an imposing figure. He was three, maybe four or five inches taller than me, intense and energetic. His bushy black eyebrows compensated for his thinning hair and focused attention on his brown eyes. He always wore either a suit or sports coat to class.

"What do you know about the scientific method?" he asked the class.

A hand went up. "You have a hypothesis and you test it," a brave student volunteered.

"Sort of," Blumberg said. "The scientific method begins with an observation, which becomes the basis of a hypothesis. Next, you design experiments to test whether the hypothesis is supported or rejected. The data generated from the experiments yield new observations, which, in turn, refine the original hypothesis or generate a new hypothesis."

"So, the scientific method is an iterative process?" the same student asked.

"Exactly. You have to be open to incorporating new information and willing to adjust your hypothesis. It is critical to think outside the accepted dogma and explore new ideas. It is also vital to reject the null hypothesis."

"Null hypothesis?"

"Testing of hypotheses often involves statistics. For example, in medical genetics, we are comparing the differences in disease susceptibility between two groups of people. We generally use 5 percent as a cutoff for the probability of these results being due to chance. If the difference between the groups is greater than 5 percent, then we accept the null hypothesis and say that there is no significant difference between these groups. We want to minimize the possibility that our findings are due to chance. Our goal is to identify differences between groups that are real and reproducible."

Blumberg was introducing us to science. This course was going to be about facts and puzzles. I wanted to learn more.

Early in the course, when Blumberg described how he had discovered the Australia antigen, he said, "I was searching for new disease-associated antigens."

"What is an antigen?" I asked.

"An antigen elicits an antibody or immune response. In fact, the term 'antigen' combines the first syllables of *anti*body and *gen*eration," he answered.

"I've spent decades traveling around the world collecting blood samples. I devised a simple way to discover new antigens in serum."

"Serum?" I asked.

"Blood is composed of two primary parts: the red blood cells that are involved in clotting, and the serum, which is the clear liquid portion left over. I used a double-diffusion method, where a cookie-cutter template is used to punch seven holes into an agar gel. The center hole is surrounded by six evenly spaced holes around the circumference. One drop of an exotic test serum sample goes into each of the outer wells and the center hole contains the serum from a person known to have lots of antibodies. The proteins from the wells diffuse through the agar. If an antigen meets an antibody, it forms a visible line."

He was describing visible, tangible, and unambiguous results. A line. Blumberg's description of his discovery mesmerized me. It was simple, and what he described sounded doable. I was hooked and wanted to know more. No cultural anthropology lecture had ever left me thirsting for more.

"Where do you find someone with antibodies? Since these samples are exotic, if you find a band, how do you know what it does?" I asked.

"Simple. We used hemophiliacs who had received multiple blood transfusions over the years as our source for antibodies.[1] We reasoned they were exposed to hundreds of foreign proteins and would have made antibodies against them. One of these experiments yielded a positive result between the serum of an Indigenous Australian and an older multiply transfused patient with hemophilia. We didn't know anything else about the antigen so we gave it a neutral name, Australia antigen, while we figured out the biology of the antigen."

"Do you know what it does?" the brave student asked.

"Our research showed that Australia antigen is associated with hepatitis B, a serious infectious disease."

"Why is it important to know what causes a disease?" I asked, wanting to hear Blumberg's answer, since my stepfather had founded the Department

---

1. The most famous carrier of the hemophilia gene was Queen Victoria. Ten of her male descendants (for three generations, i.e., among her children, grandchildren, and great-grandchildren) manifested the disease, and six of her female descendants were determined to be asymptomatic carriers of the hemophilia gene, which results in an absence of the factor IX gene necessary for blood clotting. Today there are better options than blood transfusions to treat hemophilia, including gene therapy and factor IX–specific injections. See Michael Hickey, "Why Is Hemophilia Called the Royal Disease?" *HemAware: The Bleeding Disorders Magazine*, April 28, 2023, https://hemaware.org/bleeding-disorders-z/royal-disease.

of Preventive Medicine at UCSF and my grandfather had been Director of Public Health for the State of California.[2] I had grown up with the mantra that "an ounce of prevention is worth a pound of cure."

Blumberg said, "Identification of the antigen leads to diagnostic tests—the crucial first step to developing effective treatments or prevention strategies. We know that some people become chronic carriers of the virus, but we don't know why. Are there genes that influence susceptibility? Are there long-term consequences of infection? What is the epidemiology or natural history of the disease? The more we know, the better we understand how to prevent or treat hepatitis B.

"An interesting fact is that the Australia antigen is rare in the United States but is relatively common in Asia and the South Pacific. We don't know why, but it is important to learn how some people become chronic carriers."

"Where in the South Pacific?" I asked, thinking, *This could be a dream come true.*

"The Australia antigen is endemic in the Solomon Islands," he answered.

"And the New Hebrides? The island group due south of the Solomons?" I asked.

"Nobody's studied New Hebrides yet. It could be important and interesting to study."

I was in my last semester of classes and decided to meet with him to learn more. Surely, he would have some ideas about potential research I could do for my dissertation.

I wasn't sure about the "important and interesting" aspects of studying an antigen associated with a virus that caused a nasty disease, but I liked the New Hebrides, knew how to navigate there, and it *would* be interesting to go back. John's house and setup would still be there. It would be familiar.

I'd had little interest in biology before that discussion with Blumberg, but we worked out a research topic. A year later, the Australia antigen (Au) was renamed hepatitis B surface antigen (HBsAg) because its presence in the serum indicates a person is infected with hepatitis B virus (HBV).

Trading Venues—Philadelphia for Fox Chase

Blumberg concentrated on the virus. I focused on getting back to Malo's narrow sand beaches and living in a familiar place. We knew that about 15

2. My stepfather was Salvatore Pablo Lucia, MD. My grandfather was Walter Murray Dickie, MD.

percent of people in the adjacent Solomon Islands had HBsAg, one of the highest rates in the world. New Hebrides was in the right place to examine its epidemiology. He wanted me to investigate the maternal effect—in other words, I would research whether the virus was transmitted from mother to child, and if so, how.

"Come out to the Institute for Cancer Research and we'll teach you some laboratory techniques, see how you like it. If you like it, we can sponsor you with a predoctoral fellowship," Blumberg said.

I welcomed his offer to become part of his ICR team. I would participate in their staff meetings, formulate a research plan, and learn how to do lab work to determine whether a person had an antigen or antibody to the virus. It was just what I needed. He gave me a clear idea of what I could do for a dissertation, and clear ideas were something I seriously lacked.

I started commuting forty minutes each way to ICR, located in a northeastern suburb of Philadelphia. There I learned how to test for the presence of the virus, although I had minimal interest in virology.

I participated in his staff meetings and went from lab to lab, where I learned how to do different tests to determine whether either the antigen or the antibody were present. Within three months, I reliably performed diagnostic tests for the antigen and antibody using four different types of assays.[3] These labs generated numbers, data. They were concrete, tangible, measurable findings and so different from the qualitative word-based world of cultural anthropology. It was the first time I had been exposed to real science and real biology: hard science, not social science. I liked it. It was difficult at first, like learning a new language, but once I understood how the tests worked and how the data were analyzed, it was like a breath of fresh air. Results were reproducible. What a relief. I didn't have to create a framework; the data did that. My job would be to collect the information and, with Blumberg's team, interpret it.

At that time, there was a lot of activity, excitement, and pioneering work happening in the Blumberg lab. A team of French physicians[4] working in Mali and Senegal discovered that chronic infection with Australia antigen was associated with developing primary hepatocellular carcinoma (liver cancer) later in life. At the same time, there was a collaboration with Merck

---

3. An assay is a laboratory testing method used to determine whether a specific protein or molecule is present.
4. Led by Bernard Larouzé, MD.

to develop a hepatitis B vaccine.[5] In addition, Blumberg's other anthropology graduate student, Jana, was publishing the data she'd collected in Greece on the maternal effect of virus transmission.[6]

One day, during our one-on-one weekly meeting, Blumberg broached the topic of my dissertation. He said, "We know hepatitis B is transmitted parenterally through blood-to-blood contact. Jana's Greek data show that children born to HBsAg-positive mothers are significantly more likely to become HBsAg carriers. We think this maternal effect is due to transplacental exposure to the virus and/or exposure to the mother's blood during birth. I'd like to test this hypothesis in a different population. How about studying the maternal effect in New Hebrides?"

"Sure, that sounds fine," I said. Confirming someone else's results didn't sound groundbreaking, but it did get me back to Malo, and the research sounded straightforward. If I found the maternal effect on Malo, that would confirm his hypothesis. The rationale was that early parenteral exposure came from an HBsAg-positive mother, which caused the child to become a chronic carrier of HBV. That was the prevailing wisdom at the time—and to a great extent, it's the way it's still thought about today.

Blumberg was a world traveler, supportive and keen to send others out into the field as well. He professed to love fieldwork and assumed everyone else would, too. I subsequently heard that his idea of field research was more about staying in hotels than roughing it in the bush.

He was not an anthropologist, and I've often wondered how hands-on he was as a medical geneticist. For him, I suspect, it was all very civilized. Blumberg never struck me as someone who walked from village to village, camping for months on end and worrying about water quality. In me, he found a student who needed a project for a dissertation and who wanted to return to New Hebrides, an area of the world that was scientifically interesting to him and relevant to his research. He could send me out to do research and would have access to whatever data I collected. If it confirmed his hypothesis, great. If not, it meant that there would be more new observations and new hypotheses to test. Either way, it was a win-win for him. He had discovered the virus, had a strong research team at ICR, and

---

5. The main contact was Dr. Maurice Hilleman, who led the development of the HBV vaccine. He and his group developed forty vaccines during his tenure with the company.
6. B. S. Blumberg et al., "The Relation of Infection with the Hepatitis B Agent to Primary Hepatocellular Carcinoma," *American Journal of Pathology* 81, no. 3 (1975): 669–82.

had found people who would send him samples to analyze.[7] Furthermore, his teaching position in the anthropology department provided him with access to students who wanted to go to far-flung places. Unfortunately for him, only two of us were interested in doing our doctoral research on HBV.

Around this time, Blumberg started outlining his discovery process and journey with Australia antigen. He mapped out each hypothesis, along with results and observations, on large pieces of butcher paper taped on the four walls of his office, each wall representing a different phase of his process (discovery, identification of the virus, disease association, cancer). Each piece of paper specified the original hypothesis, observations, and new hypothesis.[8] He was very conscious of the scientific method and of his legacy.

Learning a New Language—Science

ICR sponsored lectures most workday afternoons, between three thirty and four thirty, when visiting scientists would talk about their work and the science underlying their research. Through this forum, I was introduced to the vocabulary of immunology and an entirely new world of thinking. There were many times when I wondered what the lecture was about—it seemed there was no end to the types of cells involved in the immune system. Some were described alphabetically (T- and B-cells), others had descriptive names like helper or natural killer, while other names were multisyllabic with no obvious hook, like macrophage. In any specialized field, there is always a unique language. At ICR, that language was science, and I had to decipher it. It was a puzzle, and I enjoyed the challenge. In the years since, I've spent my career learning these languages by constantly asking myself, *What does that mean?*

Working at ICR was foreign to me—but, at the same time, it felt like going to New Hebrides. I was trying to figure out what was happening, where I fit, and how it all made sense. When I started, I didn't have a clue about science, immunology, virology, or epidemiology—it was a whole new world. I

---

7. Harvey J. Alter and Baruch S. Blumberg, "Further Studies on a 'New' Human Isoprecipitin System (Australia Antigen)," *Blood* 3 (1996): 297–309. Many consider Dr. Alter a co-discoverer of HBV. He subsequently won the Nobel Prize for his contributions to the discovery of the hepatitis C virus.

8. The result was his memoir, *Hepatitis B: The Hunt for a Killer Virus.*

had spent two years doing social anthropology, but now I was immersed in science, and I was learning biology. My next challenge would be to merge anthropology and biology.

I had always been attracted to the intersections between disciplines. I found these gray areas more interesting than staying within the framework of a traditional discipline. There is a reason for this. Part of it is that there's less competition at these intersections. (I didn't have to compete and lose, as I did with my younger brothers and their different fathers.) The intersections also gave me a way to assert control over my work product. (I didn't have to compete with the Phi Beta Kappa scholars who functioned within the traditional discipline boundaries—which was hopeless.) Working at the intersection was a way of rebelling against someone else's structure, ensuring I would not be tethered to someone else's vision. Whenever I struggled with Blumberg, it was always about control of my research and what I could or could not do. I hated not having a say in my destiny. But, there are many different types of control, as I was about to learn the hard way.

Lessons Learned

- If you're dissatisfied with where you are, embrace new things.
- Keep learning, keep trying, keep doing.
- If your career offers a choice between fun and not fun, always pick fun.
- Complicated science boils down to doing a bunch of simple things, like putting blood on gel and paying attention.
- The basics of the scientific method are explainable.
- Hard science produces data; soft science does not.
- Specialized fields all have their own language.
- To maintain control, don't fit inside the box.

Chapter Six:

Two Many Men and Just One Woman

I met a scientist at ICR. Marty was a tall, lean, and handsome biologist who was interested in antibodies. I was more interested in his body than in antibodies. He'd been educated locally with his PhD from Penn. Like Blumberg, he led a team of talented immunologists and was internationally known. It was complicated as he was married, but that was his issue to deal with, not mine, so we started a relationship.

I was surprised when he gave me a ticket to join him on a trip to California for a conference (where the affair began). After that, I spent more time at ICR to be with him. Marty was thoughtful and reflective. He appreciated my intelligence and patiently explained aspects of immunology. It was all very exciting for me, but looking back, I believe it was just a way of life for those men in senior positions. They saw young female staff and students as a perk of the job and had no qualms about getting involved with them. A number of these men had sweet setups with two offices: one for them, one for their associates next door.

I wasn't wed to ICR and Marty wasn't the only man in my life at that point, as there was an available pool of older, confident, tenured academic men. At Penn, there was another professor who would become a major character in my life over the next four years. Bill was an anthropologist whose area of expertise was the South Pacific, and specifically, Melanesia. He was twenty-seven years my senior and, of course, married but sort of separated from his wife. He'd just returned from a year's sabbatical. He was dynamic and charismatic with a full head of white hair and a can-do presence. He chain-smoked and had a deep, gravelly voice—all in all, a fun guy. He was a drinker, a real party animal. He had a knack for storytelling. In the coffee shop, his table was crowded as we all sat rapt while he talked about other anthropologists, how the Santa Cruz Islanders can tell if someone is actually sick or malingering, and being on a ship in Pearl Harbor. Everyone loved Bill.

"Have I told you that I ran away from home at fourteen and stowed away on a ship?"

"No," someone would respond. "Why? And where did you end up?"

"I loved the sea and living in Cucamonga was stifling. So, I snuck on board a cargo ship and weeks later found myself in Singapore. I became something of a mascot for the crew. I got into a bit of trouble there, and fortunately, my new friends got me out of jail."

"Why were you in jail?"

"Something about being drunk and disorderly . . ." Everyone nodded and chuckled at the image of the fourteen-year-old delinquent but charismatic Bill.

After Singapore, he ultimately made his way to Hawaii to ride the big waves.

"I was seventeen years old. I lied and joined the merchant marine and then the navy. I was on the *Arizona* during the attack. I was injured but was able to get off before it sunk. Rehab was hard and long." He had been awarded the Purple Heart but was unable to have children.

The September before I made my second trip to New Hebrides, Bill returned to Penn. Most evenings, at six o'clock, he and I would sit in the graduate student lounge and chat over a cup of coffee. I learned everything about him. This went on for many months until, almost inevitably, there was a party one night where we both had too much to drink. The next thing I knew, we were on the floor in his office. It developed into a casual affair; at the same time, I was also involved with Marty. It was on and off with Bill. With him, things usually happened if we were drunk in the same place at the same time. It was a busy year—1974!

I felt differently toward each of them. I was in love with Marty and loved being with Bill. My feelings for Marty were strong. I have questioned why I felt that way—why him, and why such intensity? I think it was because he was married. I was naïve and intoxicated with a sense of conquest, a belief that I had "won" him. On top of that, I knew nothing about immunology. The fact that he was a well-connected immunology rock star made him attractive. It seemed like a good idea to absorb as much "science" as I could from him.

Women's World

Anthropology wasn't an exclusively male world, and I met some remarkable women, including Margaret Mead. When I was married, she had lectured annually to the Cincinnati Psychiatry Wives Group. I decided to attend, hoping to be inspired by a woman who had made such strides in anthropology and was held up as a beacon of feminist achievement. She bragged that she came from a line of accomplished women. Then she made a mistake. She chastised us wives for fulfilling our dreams through our husbands and settling for housewifery. Wrong.

"I beg to differ," I had said. "I come from a line of accomplished women, and I'm going on to graduate school, so please don't make assumptions."

She was taken aback, and the other wives looked at me in disbelief because I had challenged her. Mead became defensive and dismissive. She went out of her way to avoid talking with me after her lecture. I don't know whether Mead thought she was motivating us. If she did, her message was lost on me.

I was intent on having a career, and starting with Marty, I was on an ego trip with my various assignations—if a man was important, it was a tick in my favor. Paradoxically, I never thought about relationships as a means of advancing in my career. Instead, I tried to avoid mixing the two. When I first slept with Bill, I was taking his class and quickly realized things could be really awkward, so I deliberately didn't finish the required final paper.

"Do it," Bill told me. "You'll get a good grade."

"Absolutely not. I'm going to take an incomplete," I replied.

It was the one course in my life I never finished. I felt compromised. I took an additional course instead, as it was not right to have that crossover, and I would always feel that my grade was tainted due to my affair with him. I wanted to keep my professional self as clean as possible, as far away from my personal life as I could manage.

Ironically, it was the women in my life who generally impacted me the most. I am aware of how much my mother's behavior affected me, from the moment she decided to take me to Europe with her when I was three years old. Soon after Mom wed her second husband, my adopted father, Walter Dickie, she experienced two profound disappointments: rejection from medical school and the birth of my brother, who was expected to die within weeks from melanoma (which was actually a benign infantile nevus). I was old enough to travel, so she took me with her, leaving Walter and his parents to cope with the infant. This was her way to avoid bonding with Terry and minimize the pain of his impending death. I also had my godmother, who provided stability and structure. She wrote award-winning mysteries and penned speeches for Robert G. Sproul, President of the University of California. I was surrounded by accomplished women.

The next woman who had a profound effect on my life worked for Blumberg.

At ICR, Blumberg's team taught me how to do the different diagnostic tests in a sophisticated lab environment. However, when I went back to Malo, I wouldn't have the same equipment or reagents. I would have to make allowances for that.

Blumberg's fourth in command, a woman named Scott Mazzur, had done hepatitis testing in the field on Santa Cruz Island in the Solomon Islands, where she'd liaised with Bill. It was Bill who connected us, not Blumberg—because Bill understood the challenges I would face in New Hebrides. The Solomons are the island group immediately north of New Hebrides, and Scott had done a summer of fieldwork there as a microbiologist taking blood, testing it, and then returning to ICR to analyze her data. I liked her immensely.

"I can teach you how to do antigen and antibody testing in the field," Scott offered. "Barry's used to doing fieldwork with a team of people—an entourage that draws blood samples. His teams have access to centrifuges and refrigerators, everything one might possibly need to preserve the sample for laboratory testing. He has no clue about what you'll need. Don't let him tell you that you'll do all the testing here in Philadelphia. You'll be much better off if you have test results while you're in the field, and you'll make better, more informed decisions."

"Okay. Tell me what to order and how to do the tests."

"I will but wait until I've left ICR. I want to have a clean break, and it looks like I'll be taking a position at the American Red Cross in Washington."

After Scott moved to Washington, I visited her and she told me how to translate sophisticated diagnostic tests into equally accurate field-based techniques. I dutifully wrote down which reagents and equipment to buy, how to pack and ship the material, and who to contact for commercially unavailable reagents.

"I'm going to return to Santa Cruz Island this coming summer, and I'll visit you on my way there," she said. "I'll teach you how to do everything from blood collection to reading test results in the field, but you'll need to have all the equipment there. It'll work. We'll make it work as long as you do the preparation."

There were so many lists—supplies I had to get, methods I had to learn. Blumberg gave me the money to buy everything. I placed my orders and learned phlebotomy (how to draw blood) at Jeanes Hospital while I waited for the equipment to arrive. No one checked what I ordered; Blumberg told me to get the supplies I wanted and pack them when they came. That's exactly what I did. I bought a trunk and loaded it with supplies, and the ICR staff shipped it to New Hebrides.

Other considerations were not quite so straightforward. The field-compatible test for hepatitis B antibody (anti-HBs) wasn't commercially available and I needed it. Scott came to the rescue. She knew someone at Abbott Laboratories who'd developed a field-compatible test (Auscell) for HBsAg. Her contact also had developed a similar test for anti-HBs, and that test was used for Abbott's internal research. She suggested I contact Lacy Overby at Abbott Laboratories and ask for the anti-HBs assay. I did exactly as she suggested. When I reached him by phone, I told him which reagent I wanted, why I wanted it, and where it would be going.

"I served in New Hebrides in World War II," Lacy told me. "I remember it well. Of course you can have the test—for free." He made the antibody test available to me, shipped it, and became an important technical resource for my research. I was thrilled. The complementary freeze-dried (or lyophilized) Abbott tests could withstand changes in heat and humidity. When I wanted to use one or both of the tests, they had to be reconstituted. Both Abbott assays were third-generation tests, so the results were clinically robust and actionable; the only difference was that Auscell was commercially available and the antibody test was not. However, translating Scott's verbal instructions into practice would involve trial-and-error learning if she didn't make it to Malo to demonstrate in person.

Meanwhile, Blumberg and I constantly disagreed about how and where to do the laboratory tests. Scott never wavered in her support or her generosity in teaching me. This was a valuable lesson. She was the first woman in my graduate career who went that extra mile for me. She saw me as someone to assist—and her help made everything that followed in my life possible.

Blumberg directed me in his usual commanding way to present what I was going to do in New Hebrides to his team at ICR. This was a trial by fire as I was so ignorant compared to the professional immunologists. I grasped at straws and did my best to explain that I would collect blood samples and observe how culture and behavior influence disease transmission. I had no idea what I would really do, and I was sure they could see that. I suspect Blumberg rationalized that he had very little to lose, since his entire investment was less than $5,000—a rounding error in his multimillion-dollar operating budget from the National Institutes of Health and the National Cancer Institute. In addition, the funding reinforced his position in the anthropology department, since he was supporting one of their graduate students.

I argued with Blumberg a lot—which would have been awkward even if my classmate hadn't been in a relationship with him. She became something of an intermediary. She relayed his frustrations about me and offered advice about how I should talk to him. Penn and ICR were rife with inappropriate behavior in the midst of the sexual freedom of the seventies.

There was still one more hurdle I needed to clear before I could return to Malo—my oral exam. I needed to demonstrate my expertise in Oceania, my geographic area, and defend my research proposal before the entire anthropology faculty. When I passed the exam, I celebrated by cooking an outrageously complicated French meal that took three days to prepare. I invited the Goodenoughs, my next-door neighbor Chet Gorman, and four student friends. It was a memorable event. We each consumed an average of one and a half bottles of wine over the four-hour meal, and afterward, Chet didn't make it back to his apartment.

There were no more administrative obstacles from Penn. This meant that Blumberg's grant paid for me to go back to Malo Island with a salary of $300 a month. I was young, I still had a lot to learn, and it was time to discover my next chapter.

Lessons Learned

- When you're romantically involved with someone, you're more motivated to listen and learn what they know.
- Conquering people is satisfying.

- Travel far away from a situation too difficult to face and leave others to handle it.
- Take advice from all sources, but make your own decisions about what best benefits you.
- To keep your professional life clean, don't mix it with your personal life.

Chapter Seven:

I Get Around

On my return to New Hebrides in 1975, I opted to go via Thailand—not exactly on the way, but my neighbor Chet, a Penn professor, was leading the excavation of Ban Chiang, site of the earliest farming and metal use in Southeast Asia. I had three reasons to visit: the tickets were incredibly cheap, I'd never been there, and I'd confirm that Chet and I were better as friends than as lovers. I was following my mantra from my island-hopping summer—if you can affordably visit a place, there is an adventure that awaits.

April 15, 1975
Dear All,[1]

The trip to Bangkok was a real bitch. Eighteen hours of straight boredom. Not really all that bad to Tokyo. But the worst was the two-and-a-quarter hours between Hong Kong and Bangkok—that time when the end seems so close, but is still hundreds of minutes and thousands of seconds away.

Sunday and Monday, I took the train up to Ban Chiang. A sleeper of my very own and a compartment to call home. Shades of the Orient Express, but Marlene Dietrich . . . I have yet to become. The village is a very wealthy one by northeastern Thai standards and when I got there they were in the midst of celebrating their New Year (Songkran). The whole day was fantastic. We walked around in a procession throwing

1. Blumberg wanted me to keep a diary, but I refused, so instead I kept carbon copies of my letters. The onionskin paper was so thin and the blue carbon so faded that they would not xerox, so I had to dictate them.

water on people, cleansing them of their sins. We were drenched as well, but it was incredible.

The housing here is on the near side of elegant. At this point I am convinced that all the pictures of Malinowski in a tent in the Trobriands were made up, and in fact he had a villa on the leeward side of the island with a house staff of twelve, electricity, and running water.[2] *Not to mention an electric typewriter and Dictaphone. Ah yes, the hard life of the anthropologist. Why don't they tell us the truth when we're going to school? It is the only time in the course of our academic training that we have the chance to live royally. Perhaps that is what cultural relativism refers to—go to the field so you can come back to your own society understanding how the upper and aristocratic classes survive.*

I plan to be here until the middle of next week and then head down to Bangkok to outfit myself in the latest and most stylish of fashions. Actually, I am anxious to arrive in New Hebrides. Really want to see the old gang again. Ah, to return to the life of the endless gin and tonics.

I do miss you all, but not too much, as it still seems like vacation and the semi-permanence of it all hasn't quite hit.

Love, Betsy

I don't recall visiting the actual dig. I spent my days watching some of the artifacts be cleaned and cataloged. On my second evening in Thailand, we went out for excitement into the city of Udon Thani, where there was a significant US Army presence. In a bar, on the television, I was exposed to my first pornographic movie. Yes, a Shetland pony had a leading role. There were also women performing on tabletops and atop laps for money. Definitely different.

Even using a toilet in Udon Thani was dicey. Here, I encountered my first Asian squat toilet. The concept was clear and I went into a stall and closed the door (there were no locks). While I was in full squat, the door opened abruptly with a GI and a young Thai woman barging in for some privacy. "Oops. This stall's mine for now," I said. The GI processed the awkward situation and then backed away with a meek "Sorry."

My three days at the site were impactful. I got dysentery but had a wonderful time seeing my friends and observing how they handled a relatively

2. Bronislaw Malinowski was a Polish British anthropologist whose ethnography, *Argonauts of the Western Pacific*, is a classic. He is recognized as one of the founders of social anthropology.

primitive living situation. We played games and bantered, each of us trying to be cleverer than the next. Someone found an article discussing the inherent sexism in the English language with the use of "cock" in so many words, like "cocktail," "cockeyed," and "stopcock." A fellow graduate student, Bill S., won the contest cold when he quietly stated, "Bangcunt is the capital of this cocktry."

April 22, 1975
Dear Caroline,[3]

The stay at Ban Chiang was fantastic. I don't think I've ever eaten so much good food in my life. Thai cooking combines the best of Chinese and Mexican food. In fact, the whole "vacation" in Thailand was swell. It was great to get accustomed gradually to both tropical and village life again.

Perhaps you're wondering why this is written so poorly. It is for two reasons:

I am more than slightly on the loaded side—Chet and Debbie [his then girlfriend] took me out to supper and I had two drinks with a Kobe beef chaser—and I am on "The Insurgent Express" heading not so rapidly for Bangkok.

Well, now down to the nitty-gritty. What the hell happened at the physical anthropology meeting in Denver? Tell all. As for my love life—what is it? I called and talked to Marty from Honolulu, but that's all I've heard from him. Granted it's a little difficult to keep track of me right now. But damn him. He did say that he would call, however, and I gave him a phone number in the New Hebrides, so we shall see.

In Bangkok I'll be spending a few days in palatial luxury in the house of the naval adviser to the Thai government. Then on to Penang, Malaysia, to live the life of Somerset Maugham for a few days and then to Port Vila. I'll be a few days late, but such is life.

Love, Betsy

3. Caroline (Connie) was a classmate and close friend. She also worked for Goodenough. She was about nine years older than me and was interested in archaeology and physical anthropology. At the time, she was divorced and raising two young boys in the Philadelphia suburbs. Like many twentysomething women at that time, we talked about and shared tons of private information about our love lives.

Chet had made all my travel arrangements. On my return to Bangkok, I stayed with Helen and Chuck. Their home was in the former residence of a Thai princess, which was located on a klong (canal). I'd met Helen in Ban Chiang, where she occasionally volunteered. Chet introduced us, and since I was a Californian like her, she invited me to stay with them. Helen was socially active and an accomplished reader of tarot cards. During this period of student unrest and the Vietnam War, it was common to seek answers about the future from unconventional sources.

April 26, 1975
Dear Marty,

I'm bombed—totally and completely high—the Black Russian did me in. However, this evening Helen read my Tarot cards (she learned how to do it in Taiwan and damn if she isn't second-sighted).

So the question I had in mind was: "Will Marty and I get married?" She said that from the cards, there was absolutely no question but that the answer was YES. So, get your act together—God, I love you. Also, Chuck read my palm and yes, he said that my heart was ruled by my head (not exactly a surprise and accurate as all get-out) and that I would lead a relatively serene life and a long one. Helen also said that my research would be routine but that it would work. Personally (and you'll love this), she warned me to stay away from Chet and I told her that in no way was I emotionally involved and that you were very much in the picture.

I have been holding up fairly well. I only think about you constantly, you know, walking down the street here and tripping, because my mind was halfway around the world, waking up at 3:00 a.m. and unsuccessfully fantasizing that you are really here, etc. Quite honestly, I miss you more than I thought I could. Funny, how extremely important you are in my life, and in terms of elapsed time, it was so short. There are moments when I seriously doubt that the intensity will ever dwindle from our relationship. Damn it, Marty, you're stuck with a permanent lover and you best get accustomed to the idea. I love you and I want to marry you. I never really thought I'd say that again, but it's true; more than anything, I want to legitimize our relationship.

It's about time for me to go take a shower and get my act together so I can see Bangkok after dark. I love you. I love you. I'll see you in California.

Betsy

On my second night in Bangkok, Chuck and Helen had a dinner party. Two natty middle-aged men showed up.

"You're a professor at UCSF?" I said when they were introduced. "That's a coincidence. My stepfather's a professor there, too."

"Who's your stepfather?" the more outgoing professor asked.

"Salvatore Lucia. He's the chair of the Department of Preventive Medicine."

There was a pause as he stared at me in my Indian cotton granny dress.

"Your mother is the other woman," he declared.

That took me aback. Actually, Sal had been long divorced when he and Mom started their affair. My eight-year-old self had not appreciated that it had been scandalous when my mother divorced my adopted father to live with her medical school professor. They married just before the birth of their first child, Espy (S. P., or Salvatore Pablo Lucia Jr.), my second brother and Sal's first child. Espy was born nine days before Sal's sixtieth birthday. Their second son, Darryl, was born the following year. We are half a generation apart; I am eleven and twelve years older than they are.

The other topic of conversation that night, April 29, 1975, was more serious: the fall of Saigon.

"We are beginning the evacuation of Americans from Saigon," Chuck said, confirming the accounts in the newspaper. "We expect the Vietcong to take control of the city any day."

"I guess that means I've lost my shipment of blue-and-white antique porcelain," Helen said, personalizing the impact of the news. She had magnificent antiques on display in their home but wouldn't be adding to her collection.

"Does this mean it's the end of the Vietnam War?" I asked. "What happens next? Will Thailand fall next?"

No one knew the answers, but I was glad to be leaving the following morning. There were many questions and much discussion about the evacuation of Saigon and whether Thailand would be the next "domino" as Communism infiltrated Southeast Asia. This moment ensured I would remember where I was at the end of the war that defined my college years.

"You'll Be Fine"

I caught a plane to Penang the next day, just like Chet had suggested. As Peter had been in Fiji, Chet was my travel muse. He was filled with suggestions for where to stop, where to stay, what to eat and drink. I followed his advice without regret and with some important consequences. On arrival, I went directly to the Eastern & Oriental Hotel. I asked for W. Somerset Maugham's room and was given the key to Room One in the Old Wing.[4] At a nearby bookstore, I purchased one of his books and began reading *Of Human Bondage*. I didn't know at the time how appropriate the title was or how relevant its contents would be.

I hiked up Penang Hill and saw the Indian Ocean for the first time. I ate satay (grilled meat served on a skewer) and my first street food with abandon. I also bought a batik sarong and ikat weavings. I dined alone in the hotel's restaurant.

After taking my order, the twentysomething-year-old Malaysian waiter asked, "What brings a lovely woman like you to Penang?"

Naïvely, I said, "I'm an anthropologist passing through on my way to New Hebrides."

"When you are finished, I can show you some sites the tourists never see. Come with me."

"No thank you. I have work to do after dinner."

That first rejection wasn't enough, and he kept trying to make it more attractive by including a night on the town with his friends. This was one adventure I turned down. I was unnerved by the waiter's suggestion and the creepy feeling it gave me. I was afraid he might follow me when I left the restaurant. Thinking it would be good not to be alone, I took a cue from Hollywood, walked over to the table of another single diner, and asked whether I could bum a cigarette.

"Oh, please do join me," he said.

He was friendly and that was all the further it went, but we were being watched. The waiter kept his eye on us, and when we left together, he gave up.

My next stop was Singapore. The plane took on passengers in Kuala Lumpur, where a middle-aged Norwegian businessman boarded and became my seatmate. He and I both had an eight-hour layover in Singapore. He offered to show me around.

---

4. A one-night stay would now cost around $250 in 2024. It was much cheaper then!

"Sure, that sounds great. But there are a couple of things I have to do. I need to buy a camera and I have to have a Singapore Sling in the Raffles Hotel."

"Not a problem," he said. "Camera first. Let's go to the Thieves Market."

Sounded good to me!

We took a taxi to the Thieves Market and then trishaws around the city, visiting the sights together, courtesy of what I assumed was his expense account. As we wandered around the Thieves Market, he pointed out which merchandise was good and which wasn't. Increasingly, I valued and trusted his judgment. On his advice, I bought two Nikkormat cameras for my fieldwork along with three lenses, which depleted my bank account. Over *the* cocktail in the Raffles Hotel bar, I shared my fears with this experienced and patient man.

"I'm scared of being by myself and alone for two years in New Hebrides. I've never done independent research and have never even tried the laboratory field techniques. I keep worrying that I'll fail and will have spent two years of my life with nothing to show for it."

"You'll be fine," he calmly told me.

"How do you know?" I was shocked that a complete stranger could be so sure about something so important to me.

"I can tell. I know people. As a salesman, I work with lots of different people," he replied. "You're resourceful and smart. You'll figure it out."

I couldn't just accept that. "How can you tell? How can you *know* I'll be fine?"

"Really. You'll be fine."

His certainty was comforting, and his words never left me. In my darkest moments, I replayed them. This stranger knew my capabilities better than I did, and he had more confidence in me than I had in myself.

At the airport, we parted ways. I continued on to Jakarta and changed planes for the long haul to Nouméa, New Caledonia. I was getting closer. I hadn't expected the flight to be so long, since Indonesia and Melanesia were in the same "-esia" part of the world. The one-hour flight from Nouméa to Port Vila, New Hebrides, had me in a cold sweat as the reality of the months ahead became increasingly real—minute by minute. I would be living for the next two years on my own with no one to pick up the pieces if I fell. I would be doing research I was minimally prepared to do. I was going to study a virus that could make me sick.

That flight from New Caledonia to New Hebrides was the longest, most terror-filled hour of my life. Midair, I realized how truly incompetent, immature, and unprepared I was. I wasn't ready to bring all the pieces of my research together: collecting blood, making observations about child-rearing, quantifying behavior and somehow tying it to virus transmission. Plus, there were the immediate logistics to sort out: find my trunk, get it to Santo and then Malo, set up house—the list was endless. All the proverbial shit hit my psychological fan. I would have to do all of this. I sat looking out the window with an internal voice telling me, *There are no excuses now. This has to happen. You're twenty-six, you're going there for two years, and there is no turning back.* I had a visceral sense of being alone and having my fate in my own hands, yet I had no idea what exactly to do. If I failed, I had no marketable skills. It wasn't clear whether I could find another dissertation topic, but I didn't want to find out. I did not want to return to Philadelphia as a failure.

As I disembarked, I looked out of the window. I picked up my red backpack and duffle bag. The Norwegian's words finally made themselves louder than the fear pounding through me. *I'm going to be okay. I'll be fine.* I hoped.

The hope wasn't without difficulties. I was still madly in love with Marty in America, and my letters show a Betsy who was constantly wailing, *I love you! Why aren't you writing back to me?* I was pouring my heart out and getting nothing in return.

A Quilt of Coincidences

In some respects, I had as little control of my work as I did in my relationship with Marty. Everything—Marty, my research, even lodging in New Hebrides—was all based on hope. I didn't have an itinerary or any reservations—you didn't for this sort of trip back then. You just found a place once you arrived. I had to get my equipment from Vila to Santo, which was 170 miles by sea, but, again, I had no actual plan for how that would happen. I knew some people who had moved down to Vila, like the Weightmans, the couple whose party I'd attended the last time I was "in town" (on Malekula). Barry was now the head of agriculture, so that was a start. If I called them, I knew they would definitely ask me to dinner, and I could stay with them if I needed to.

I needed to.

Then, someone told me the British district agent (BDA) for the Northern District (including Espiritu Santo and Malo) was in town. Perhaps I should talk to him. I did—and he was very helpful.

"My boat, the *Rosinante*, is here. Bring your trunk to the wharf today," he offered, "and I'll transport it to Santo for you."

I collected the trunk with all my medical supplies and my boxes, which contained even more stuff, then assumed that, somehow, I would be able to get it all from the dock to his boat. An unspecified "someone" would have to offer to help carry my cargo.

It will happen, I told myself. It would work out, one step at a time.

I had no grand plan. Instead, I followed lines of suggestion and notions of who might be interesting or useful, and everything came together. I made connections, and my blind faith in talking to people and accepting help from strangers ended up creating an amazingly rich network—a coherent "quilt of coincidences"—out of what at first seemed to be nothing but patchwork.

So when the BDA told me I should meet someone from the World Health Organization (WHO) whom he knew, I simply said, "Okay!" That's how I found my way to the WHO office and asked to see the person the BDA had referred me to.

"What is your project? What will you be doing?" the Cuban WHO physician asked me—one complete stranger to another.

"I'm going to be taking blood samples to test for hepatitis B," I replied.

"You know, it would be really helpful if you could save a drop of blood for malaria testing," he said.

I was open to everything. "How do I do that?"

"Well, let me show you. You take a slide," he said, giving me a box of them, using one to demonstrate. "Just put a drop of blood here, then label and number it and make sure we know who the person is. Simple as that."

"But it's from a finger stick," I protested. "How do I take a finger stick?" I'd only been taught to do a venipuncture. This was new to me.

"You just prick the finger with this lancet," he said, giving me a box of them. "It's so easy, and this would be such a nice thing for you to do, since no one has ever collected blood samples on Malo. If someone's positive, we can treat them. Works out for everyone."

He then instructed me on how to send the slides back to him. He was happy, I was happy. He was getting tests that he wanted, and I could help the islanders.

I was beginning to see that the Norwegian might have been right. I was following leads, going from place to place and person to person, and it was working. I listened, I acted, I followed through. My equipment and I safely made it to Santo, and I made use of Ken Hutton's mailing address.

Piece by piece, things started to fall into place.

Daddy Issues

There was only one piece that did not fall into place once I arrived in New Hebrides.

When I first wrote to Marty, my letters to him at ICR were in plain white envelopes with single nondescript airmail stamps. When I didn't hear back, I decided to take a different approach. My letters were now covered with exotic, beautiful stamps—the kind everybody notices. Finally, in Santo, I received a communication from him. The telegram was three pasted paper strips with its message in capital letters: "ADDRESS NO GOOD. LETTER FOLLOWS. MARTIN."

I was thrilled—a letter! Weeks passed as I waited. The next letter I sent was even more desperate.

May 31, 1975
Dear Martin,

What the hell is going on? This whole thing is driving me up the wall. "Letter follows . . ." What letter?

For God's sake would you please write and tell me something? At this point I almost don't care what you say, but just say something! This is really absurd. I'm a reasonably patient person, but right now I'm a lot on the depressed side. I'm tired and I'm probably sick. There are so many conflicting feelings, thoughts, and actions that are within me right now. I'm beginning to think that my only option is to withdraw to Malo. Totally and completely immerse myself in nothing other than my work.

I guess what I want to know are answers to such questions as: "Do you love me? Will I see you in January?" Or, if everything was just fun and games, for God's sake tell me. I have to know where things stand. Right now I am feeling very lonely, and my self-image isn't all that hot either. I mean, there are people around and I could go out and party, etc. But right now, I'm tired of all that, I'm tired of superficial relationships and conversation. What I want is something permanent, something that has a foundation and that can be built upon and developed. I'm tired of being cute and witty and the life of the party.

Damn it, I want to be a recluse, the shy and quiet type. But most of all, I want the security and comfort of something permanent. I want to be able to give all I can and bask in receiving a similar love. I don't want to kid around with ifs and maybes or perhapses. I've given to you as much as I could and can and would just like to know if I should continue to do so. I can't live on daydreams and fantasies for seven to eight months or for sixteen months. I have to know where things stand.

I think all I really want out of life right now is to be able to do my work, but more than that, I want to have a permanent and meaningful relationship with a man. I would prefer that man to be you.

Betsy

Finally, after no response, even I realized it was over and was time to move on.

Thursday, June 10, 1975

Dear Martin,

I hope that this letter finds you well. Sorry to violate your wishes about sending a letter to ICR, but I'm competitive and prefer closure when possible. As I haven't heard from you, and it doesn't seem likely I will hear from you, I have inevitably reached a number of fairly obvious conclusions. Right now it doesn't upset me that I was taken for and played the fool extraordinaire. What does bother me is your lack of common courtesy to tell me the rules had changed.

Yet, if the above isn't true, I could have used some moral support during the last three and a half months. I think this aspect is character revealing and here is where the damage may be irreparable.

I still plan to spend a month in California at Christmas with my friends and family. If you think we have things to discuss you can reach me at home or write.

Thank you for three good months. I'm sorry the aftermath was less than the promises. Yes, you were right, things may be different when I return.

Betsy

I was firmly in the middle of what I think of as my "narcissism" phase, which meant I was attracted to accomplished, self-centered men who were like my stepfather. I basked in their reflected glory and the status they had. It wasn't only scientists. There was also Remo, who was an artist/engineer. I thought I was becoming a star by dating these prominent men, believing it was a feather in my cap to be having an affair with Marty, Bill, or Remo. These were big names, not random people—they were "A" players. I didn't stalk them, and I didn't target them. I just thought Bill was amazing and cool, Remo was smart and creative, and Marty was strong and handsome—and it just so happened that I ended up sleeping with all of them.

This phase seems understandable. I grew up in a complex and dynamic family situation. My mother looked like Ingrid Bergman in *Casablanca*. She was a smart, witty, ambitious, and charming host who attracted wealthy, high-profile suitors. I was neither a great beauty nor "arm candy." Subconsciously, I was seeing if I, too, could attract famous high-profile paramours, even recapitulating the twenty-seven-year difference between her and Sal with Bill. At the time, it seemed perfectly reasonable.

How did I get there? Daddy issues? I'd had three fathers by then, with Sal being the most stable. An oddly competitive relationship with my mother? She showed me it was okay to go after whatever or whomever you wanted. Lack of a stable home? We moved whenever a new brother was born. Sexual abuse? My mother's refusal, when I was five years old, to heed my pleas not to be left alone with the houseboy? She fired him only after our housekeeper showed her my bloody underwear. My mother's approach to marriage and divorce seemed clinical rather than emotional. She often said, "Falling in love is easy, so why not fall in love with a rich man?" There were so many reasons for the state of my love life; a Freudian psychiatrist, like my mother, could have a field day.

Before I'd totally written off Marty in June, I found some diversions in Santo that made it easier to move on from my obsession with him.

May 10, 1975
Dear Caroline,

I just got your letter, and I am so jazzed and excited I can hardly type straight. In fact, my fingers are shaking, it is so damn good to hear from you. In fact (not to use that phrase again), it is so damn good to hear from someone back home!

Yes, to date, my mother wrote after I called her from Singapore. Voilà, you stand alone in a privileged circle. The time away has done my soul a lot of good, and I am already a lot calmer and much more together. So far, I have been to Malo once and engaged Vira (John's ex-foreman) to be my general handyman to renovate the house, and Jole, who will teach me language, and then a house girl. I, a mere graduate student, am an employer of three. Such is the trauma, but now I must learn administrative skills—horror.

For the past few days, I have been having a severe case of the domestics. You know the syndrome, when you go into stores with visions of sofas dancing in your head. If I were younger and hadn't already been married, I would have thought that I was registering my china and stainless patterns at the local Chinese stores. But it has been grand. I've spent some of my happy hours sewing and covering six pillows that will be the back of my sofa and covering mattresses that will be the seat. I've also spent some of my time reading one of the classics—Of Human Bondage. I felt I owed that much to Somerset Maugham, as I spent my three days in Penang, Malaysia, in his room. I loved it.

I am glad to hear that Barry was elected to the National Academy of Sciences. So, I really have mellowed. It is so damn good to be out of that ICR atmosphere. I was really starting to feel stifled. I guess it is always a give-and-take. Here I am not stifled, but then again, I don't have the same kind of companionship from my friends that I have always found to be vital to my existence.

I suppose that I'll make new friends, and to a certain extent I have already, but boy am I the odd one out. More and more there are times when I want to be serious and introspective, and the social demand here is to drink a lot and to be cute and witty. There are times when this is okay, but usually I don't have the energy or the desire to go and play cutesy.

I'm eager to get all of my things to Malo as soon as possible, see the medical people I have to, and get started. There are so many small details that have to be worked out, but I would rather make my own mistakes and muddle my way through than have to clear every detail with the "Blessed."[5]

Incidentally, there was never any attempt to make it a secret that I was going to Thailand first.[6] *I had made that decision in December,*

5. This is a reference to Blumberg. His first name, Baruch, means "blessed" in Hebrew.
6. Connie had written that Blumberg was surprised I arrived in New Hebrides so late (May). He thought I had deceived him by spending time in Thailand. Since I had paid for my ticket, I didn't see why it mattered if I arrived a week or two later than my original plan.

and certainly made it quite public. In fact, I know I mentioned it to BSB and said something about my ticket being written to Bangkok.[7] *It is crazy, but I know that he had to have known. One thing that you can do for me is to oversee what has happened to my National Science Foundation application. I sure as hell could use the money. The cost of living has gone up, and it is now $20 round trip from Malo to Santo instead of $8. Also, if BSB is going to lower my salary by half, I will need the extra income from the grant.*[8] *Theoretically, BSB could write a letter of recommendation on my behalf, but I am such a chickenshit and am scared to death to ask him. Perhaps if you can convey this, he will be more mellow on my behalf. I haven't grown up entirely, give me a little longer.*

And now to the nitty-gritty from this end. Wednesday night, I was in Santo, staying at Ken's empty flat as always. Bored, I went to a dance at Mao's with Brian, a VSO who lives next door. He's British and dull. Well, he and these two female friends of his from Vila invited me, the ugly stepsister, to go with them to the dance. Thank God I did. There, at the table where we were sitting, was what has got to be one of the best-looking men I have ever seen in my life. I mean, classically handsome. Too good to be true. Well, we got to talking, and he is very quick and sharp. Clever beyond what one might ever hope to expect in Santo. So, we got to talking more and more and dancing more and more, and soon left the company of my acquaintances and moved to a more interesting table where there was champagne flowing.

After the dancing ended, I wanted to sober up, and so we stayed at the bar and I had a few straight tonics, and we got to talking about my favorite of favorite subjects—cowboys. It just so happens that his family have been horse breeders in Australia for generations, and not only does he ride in rodeos, but also he can hunt and has shown extensively in dressage. We had some common ground to say the least, and we must have pursued that subject for quite some time. It turns out that he arrived in Santo the day before I did and will be here for at least two years managing somebody's plantation. So, we seem to have something else in common.

7. BSB are the initials for Baruch Samuel Blumberg.

8. Connie's letter relates that Blumberg was considering cutting my $300 stipend to $150 because of my "disobedience."

Anyway, to make a long story short, he is incredibly nice, warm, open, and friendly. He walked me home, and unfortunately had to catch a plane the next day to go to Vila and then to Tanna. But he will be back on Wednesday, and I think I shall go to the airport to meet his plane. I don't know which one, but I think I shall be there with bells on. Stop speculating . . . the answer is "yes!" I have absolutely no willpower, and the bastard hasn't written. It is a little difficult relating to a vacuum. Anyway, I really hope that it works out, if nothing else, just for company and companionship. For in all truthfulness, there are very few people here that I would want to spend much time with, much less want to confide in.

Take care of yourself and give each of your lovely boys a big kiss from me and give Ward a quick hug.

Love, Betsy

At this time, the beginning of women's liberation, casual sex was abundant and casual affairs were routine. I played my part.

Lessons Learned

- Take challenges and opportunities as they come—you'll never walk this path again.
- Agree for the fun of it—take suggestions from people who know more about certain things than you do. You never know what you'll find!
- Listen to your instincts, especially on safety—they're there for a reason.
- Optimism can be a wonderful thing.
- If you lay everything on the line, don't be surprised when it doesn't work out.
- Hearing nothing is often the same as hearing "no." Learn from nonproductive interactions and a lack of responses, and move on.
- It's good to try to make your life a little less painful.
- Make connections and networks, and let them work for you.

- Saying "Okay!" and smiling is a good thing—you can always change your mind, but you can't always get back what you lost by being negative in the first place.
- Evidence goes further than words when trying to get one's parent to take action.
- Writing letters is cathartic.

Chapter Eight:

From Crushing Blow to Gushing Flow

When I returned to Malo in mid-May, John's house was still standing. However, my hut had disintegrated and the outhouse and shower needed to be rebuilt. Everything was in a state of disrepair. I kicked out the chickens and pigs from the kitchen and refurbished the house back into the two-bedroom cottage John had built. Back and forth to Santo I went and shopped at every Chinese store in town. I'd get a piece of foam and some material, then hand sew a cover, making myself a nice little sitting area in the living room. I bought Coleman and Aladdin kerosene lanterns (not a

FIGURE 23: Outhouse and shower. The outhouse with the adjacent shower stall is on the right. The two hundred–gallon water-collecting tank is in the background catching rainwater from the "tin" roof over the kitchen.

generator), a two-burner propane stove with a tiny oven, and a kerosene refrigerator. Actually, I bought two refrigerators, one for me and one for blood samples. I also needed a shower bucket, so I found one of those as well. I had a little lab built and even labeled it "*Tambu*," or taboo, just in case anyone didn't get the message: Stay out! (See Color Plate 5: Dining room and office.)

FIGURE 24: Main house (right) and kitchen (left). Dead coral surrounds the structures.

FIGURE 25: Kitchen sink, storage, and stove

I felt very comfortable in John's dwellings. I knew what to do, and it gave me a sense of security. I knew the lay of the land, which made it less intimidating. I also knew Vomule and Jole, my neighbors and landlord. This meant, in turn (as with everything), there was a connection to others. As long as I knew people, there was help. I still had my monthly $300 salary, which meant I could afford to hire help. Jole became my tutor to help me learn the language, and Vira was my assistant and guide. I also hired Vira's daughter, Vomoli Havo, to do laundry and housework. All for $105 a month—not so bad. I had no idea whether this was a good deal or I was overpaying. It seemed fair, and it left me with enough money to buy whatever else I needed.

FIGURE 26: Vira Joseph—all-around helper

FIGURE 27: Vomoli Havo—housekeeper. Vira and Voligo's teenage daughter.

FIGURE 28: Jole Antas—landlord and tutor

Whenever I was in Santo, I stayed in Ken's empty apartment. It was next to the one where he lived, and he became my surrogate father. At one point, a drunk person came to the flat and harassed me. I could tell from his voice that he was a local New Hebridean.

May 10, 1975—Written to myself

I am scared totally shitless. There is someone outside of my window and he has been there for quite some time now.

First, he came up and he asked me if I wanted to go to a movie with him. I had just finished going to the bathroom and had approached the mirror.

"No," I said. "I'm not interested."

I then went into the main room, and then he came to that window and asked me again if I wanted to go out this evening.

"No," I replied.

He then asked me if I was married.

"Yes," I said, "to a man in the States."

I don't think that he believed me, but I don't really care. I thought he would go away then, but he didn't. He then asked me why I didn't want to go out, and I said that I had some typing to do, and then I typed a letter to Evelyn [a classmate] and hoped against hope that Ken would come back. After I finished the letter to Ev, I was just quietly sitting here reflecting over the letters I have written when I heard a sound at the window. I can't really close the curtains, because the whole [curtain rod] will come down, and I only have the light of my Coleman lantern to go by, and so I seem to be stuck here by the typewriter, pounding out nonsensical things until Ken comes back.

Jesus Christ, I hope he comes home soon. Right now I am scared to leave the apartment. I have locked both of the doors, but this doesn't give me much solace.

Perhaps I shouldn't have gone to Mao's and danced the way I did. I really don't feel that I have done anything outrageous or anything to attract this man. Perhaps he is drunk from the dance up at the Seven Stars and maybe he is moved to greater courage by the [Nagriamel] meeting this afternoon.[1] *I just wish he'd leave. God, Ken come home. I just now looked up. I know he's there. It is going on 11 o'clock now, and I'm getting*

1. Nagriamel is a political party.

quite tired, but I'm scared to stop typing until Ken comes back, or Brian, or someone.[2]

The last thing in the world I want is to be raped, and this isn't exactly the opportune time for something like this to occur, what with just starting to work here and having to live here for the next eighteen to twenty-four months.

Right now there is a drum beating in the distance and it is approaching. Dogs are barking from all corners, and the hollow, syncopated beat of the drum continues. There is a metal accompaniment to the drum and there are many people. It is much nearer. I can hear the chanting voices and again the dogs, not just barking in solitary but braying and yelping in unison and in complement.

The drums have stopped. Now the loudest sound comes from my lantern. I wish the lights were on and that there was electricity. The lamp is fading and faltering. I think that I will go and lie down on the bed, with the light on, and keep my clothes on. If I hear Ken come home and I am still up, I will ask him if I can sleep in his spare room, or maybe I will be able to go to sleep and pretend that this never happened.

I think that I'm beginning to get a bit on the paranoid side, but nothing like this has ever happened to me before. I'm also sure that part of the fear may be because he is Black. The color of New Hebrideans' skin has never bothered me before, but then again I have never felt threatened by one either. Ken, come home! Somebody, come home. Unfortunately, Ken's is the only car I can recognize by ear. Brian's would be a guess, and I think that Brian might prove to be useless as well. To hell with brushing my teeth and washing my face. So much for the shower and shampoo.

Shit, I wish right now I were back in Philadelphia, where I knew what the problems were and where I was on the second floor. I am really getting tired now. My head hurts, my fingers ache, and my body is still. All I want is a good night's sleep and not to have to hassle with all this misery. I'm going to give the bed a try now, with the light on. Hopefully it will last until Ken gets home. Well, here's trying . . .

There were so many questions running through my mind—had I locked the door, was it secure, and how would I protect myself if it wasn't?—but at the heart of it was something deeper and darker. If the man did break in, and

2. Brian, the dull British VSO neighbor with whom I'd previously gone to the dance at Mao's.

if he did attack me, how would that be perceived? How would I—a single woman on my own in that part of the world, a woman who was independent, who drank when she went into town, who was open and talked to men—be seen? It is true that he would have been seen as despicable for attacking a white woman, but I felt so vulnerable. I was shaken by it although nothing actually happened, and eventually the man left.

I realized how vulnerable I was, but this was Santo. There were lots of people in Santo and the doors had locks, even though they were rarely used. On Malo, there were no locks. I had a machete, but that wouldn't have helped, as it was big and I didn't know how to wield it. I wondered whether this night in Santo would change how I felt in my house on Malo. Luckily, it didn't. On Malo, I was a known entity and a guest. I was relieved to return to Malo, where my safety was never in doubt. After that, I never felt completely secure in Santo and rarely went out after dark alone.

Of Human Bondage

Letters were how I shared my insights and feelings. On finishing *Of Human Bondage*, which I'd started reading in Penang but didn't finish until I was settled in on Malo, the only person I trusted with my emotional insights was Michael. We'd met at a Penn anthro party. He was my age but an undergraduate. We had spent an hour or two talking over coffee. He was a principled person who had been a conscientious objector to the Vietnam War. Despite the fact he was a handsome, talented, artistic man, I soon realized I wasn't interested in anything other than a casual friendship. At that point, I distanced myself from him. Regardless, he made it clear that he had a crush on me and didn't care that it was unrequited.

Wednesday, June 4, 1975
Dear Michael,

How are you? Congratulations on your acceptance to the University of Toronto. From what I gather from between the lines, things seem to be going well. I am glad. I suppose you are wondering why in the hell I am writing to you of my own initiative? To be quite truthful, I'm not really sure. Funny, there are things I want to get off of my mind by sharing them with someone and you seemed to be the obvious choice. I mean, they're silly things, and they're quite personal feelings. You always allowed me to talk and had constructive things to say, albeit they sometimes made me feel quite uncomfortable and forced me to reconsider what the hell was going on, and yes, reevaluate my reality testing. I guess that this is enough of a prologue.

To begin with, there are at least three things that are involved in my mood at the moment. I shall take them as best I can in chronological order.

A passage from Of Human Bondage:

It seemed to him that all his life he had followed the ideals that other people, by their words or their writings, had instilled into him, and never the desires of his own heart. Always his course had been swayed by what he thought he should do and never by what he wanted with his whole soul to do. He put all that aside now with a gesture of impatience. He had lived always in the future, and the present always, always had slipped through his fingers. His ideals? He thought of his desire to make a life: had he not seen also that the simplest pattern, that in which a man was born, worked, married, had children, and died, was likewise the most perfect? It might be that to surrender to happiness was to accept defeat, but it was a defeat better than many victories.

Now, that hit home. It is the kind of book in which each line has something to say, but the last monologue was too much. I finished the book while still on Malo and then went to Santo the next day, whereupon enter item two.

A letter from my mother with the following:

"I hope you are enjoying yourself more. Do you miss your boyfriend, the immunologist, very much? I guess you will make other attachments out there—still seems like you'd be better off with someone who hasn't already demonstrated their psychosocial future, but it is your life and ultimately you have to do what is best for you. If it can be a little less painful, it would be good. You are a beautiful, intelligent, thoughtful girl who ought to be able to get the best out of life—this does not mean you should marry for narcissistic reasons; that is, don't marry the smartest, handsomest, etc.—a

narcissistic choice only. Marry someone who loves you and will be good to you and not someone to fan your ego and impress your family and friends. Don't, don't, don't choose an egomaniac—be sure the person can relate to people, likes people, likes children, and loves you. A "catch" will probably be a real headache who'll drag and soak up all you can give while giving you very little.

This fellow sounds like he could be like that, so you'd better give it plenty of thought, as I am sure you are doing. I still think you would do well, when you get back, to think about (psycho) analysis, mostly because you seem too self-sacrificing at times and think too little of yourself—and if you married this fellow with two daughters (and you younger than he), you and the two girls would forever be in competition, a hard enough thing when the girls are your own!"

Now, indeed, I had already been giving this plenty of thought. And can see the validity of her argument, and in my most objective moments I realize that she is right. But it is difficult to get my self-image out of the doldrums and stop trying to please everyone in the world.

The third thing that entered into it had to do with being in Santo. You have to understand Santo. It is a town of about four hundred Europeans and three thousand natives (or so I'm told). The Europeans are French and English, then there are Australians, New Zealanders, Americans (very few), and a few Japanese, Tahitians, Vietnamese, etc. A real grab bag. The other thing is that everyone who's here is here for some reason—usually, they're escaping something. The name of the game becomes very quickly "don't delve too deeply" and "don't, God forbid, try to go beneath the Bar Façade."

I guess I finally just got fed up with Santo. The only way to meet new people is to go to the one bar and drink. Now, it takes all of my courage to walk into a bar by myself, not knowing a soul, and then make friends. I hate it. Then, when someone I know shows up, they're on the make. In a town filled with single and married males, there are more on the make than I could have imagined. Dull. I get to the point where all I want is to be left alone. I also get very tired of the personality that I have to affect. It is all right for a while, but after the third week it loses something.

> *Combine all of this, and what you get is a lot of self-doubt and one good solid depression.*
>
> *Now, there is one other factor involved. It seems that every time I come here [and leave my comfort zone], I get a real nesting urge. I always seem to decide that it is time for me to get married, finish my degree, and have some kids. I'm not sure if this comes from some latent feelings of loneliness—because I haven't really felt alone yet, or at least nothing I couldn't handle—or if it is a result of being tired of the impermanence of everything. I think it is the latter. The more I think about it, the less I want additional complications. It sure isn't a 1970s approach.*
>
> *I'm not really sure why I am telling you all of this. I know that I have to get this whole thing of the last week off of my chest, and this is the catharsis. All I know is that I can talk more honestly with you than with others. I guess I feel that I have nothing to pretend, and that you won't like me less if you know some of my feelings. I think that's why, at least.*

When I'd had that confrontation with Margaret Mead, I had realized strongly that I was *not* just going to be a psychiatrist's wife. That was not me in any way, shape, or form, but . . . did I want to get married and have kids? Did I?

The Somerset Maugham book was stunningly meaningful—hauntingly so, as if his character Philip's journey somehow mirrored my own—and yet I was still lost. Where would I end up? What would I choose? It was all about family versus career, as it had been for so many women for so long. I kept coming back to the same question: What was this human bondage?

Asking Questions

The house refurb took a month, and I was uneasy throughout the whole period. I was keen for Scott to come so that I could learn everything that would let me start my work in earnest. Until then, I felt like I was killing time. I had created a demographic questionnaire a professor thought could provide useful information for my dissertation or future publications. I decided to start gathering the information.

One night, during my language lesson with Jole, I remembered it was always good to get the local perspective first. So I brought out the questionnaire.

"Jole, I'd like to go over these questions with you to see if they're okay to ask." I put the paper questionnaire on the kitchen table so we could both read it even though we were seated on opposite sides.

"Age," I began.

"That's a problem for the older people. We know whether we're older or younger than someone. Some of us know when we were born but many people do not."

"Okay. Age is an estimate. What about date of birth?" I asked.

"Like age, we can make a reasonable guess but there are no good records for the older people and people of my generation. For the younger children, we have some records for the year but not the date."

"Place of birth?"

"We know where everyone was born and their place of affiliation."

"Number of children?"

"We know the number of living children and often the ones who have died as well."

"Age of menarche?"

"We don't know," Jole told me.

"How can one not know when a girl has her first period?"

Jole's reply caught me off guard. He told me Malo traditionally had the custom of a "bride price," where the husband's family gives the bride's family things of value when she joins their family. (This is still practiced today.) However, in the old days, girls were betrothed around the age of three to an older man. The girl would then grow up in his house.

I persisted. "Okay, so what does that have to do with not knowing the age of onset of menses? How can you not know?"

"Well," he said, "most women do not menstruate."

"How is that possible?"

"Many women will never bleed."

"But the women are fertile. They have lots of children and at a young age . . ." That was the answer. "Is it because they are having intercourse with their husband as they reach puberty?"

"Yes. It was the custom," he replied.

I remembered that puberty generally precedes the onset of menarche and that breastfeeding stops ovulation for months after childbirth.

"How many children would a woman have?" I asked.

"As many as they could. Twelve or fourteen was not uncommon, because only a few survived. Also, a man might inherit other wives from a brother who died or an uncle. It was not uncommon for a man to have many wives."

I thought I understood what happened. Throughout their reproductive lives, women were either pregnant or breastfeeding, thus they were in a state of perpetual amenorrhea. It was easy to imagine how a woman could go through her entire life without ever having a period, so they truly had no clue about the age of menarche.

"Most of the old women were married young. It was our custom. We don't follow this custom any longer," Jole said. "Today, girls grow up in their parents' homes and marry much later. Typically, a girl will be thirteen or fourteen or even older when she marries. The man she marries is the same age. If a girl is not married by the age of twenty, she is considered a spinster."

"What happens if a woman can't have children or if she dies in childbirth?" I asked.

"If a woman is barren, then the man is free to take another wife. If she dies in childbirth, the same. Also, a man could have other wives while he waits for the younger girl to mature."

"What happens if the husband dies?"

"Simple. His widows would be 'given' to one of his brothers. The bride price has already been paid."

In other words, women would stay in the patriline as they had already been "bought." Marriages were proscriptive, with clear rules about whom one could not marry; in the past, marriages had been arranged to maximize land rights. Romantic love was not sufficient reason to marry, but it did happen.

I had one more question to ask. "What happens if the couple gets a divorce?"

"Naturally, the children stay with the father," Jole said.

"In America, the children usually stay with the mother," I said.

"Here the father's family paid for the wife, and the children stay with him on his land."

After my questionnaire debacle, Jole suggested I go to the dispensary where the medical and family records were kept. Off I went one morning to scope out the dispensary in Avunatari Village, where I met Thompson, Malo's *dresser* (paramedic).

"Can I come and watch you work?" I asked.

"Yes."

"What does a dresser do?"

"I give people medicine if they're sick. I give them nivaquine for malaria, bandage cuts, and clean ulcerated wounds. Children come for vaccines. Maybe you want to look at the records I have on different people?" he offered, and on my first visit.

"Okay!" I replied, as always. "Maybe that *is* a good idea!"

The next day, I started filling out the pages on which I had printed my Penn-conceived questionnaire. As soon as I read the first of Thompson's records, it was clear he had a different system from mine. I modified my categories and copied his information, but I had questions.

"How do you know if this is a woman or a man?" I asked. I had looked but couldn't find where this most basic medical information was recorded. I was stymied. His cards included names and ages but not gender.

He laughed. "Well, all women's names begin with Vo!" It was simple for him, and I had a lot to learn.

At the start of each questionnaire, I'd note whether the individual was a man or a woman. If it was a child, I would note the name of their parents. Soon, by copying Thompson's records and constructing family units, I developed a sense of family size and the number of villages on the island. Thompson would talk about individuals and describe where the villages were located. His records, however, were sketchy and incomplete.[3]

I quickly realized how little I had absorbed from my first trip. In fact, I hadn't even known Thompson existed. I had learned some things about stones and pottery and even less about traditional customs. In retrospect, the three months I'd spent on Malo two years earlier had been little more than a holiday.

On my own now, I had to take the initiative and jump into the culture using whatever I could. I'd started by doing something so simple and basic, and I ended up learning about naming, family structure, and the different villages. I was starting to feel pretty good and had a semblance of a daily routine.

Now all I had to do was learn how to collect and analyze blood. I wondered whether Scott would actually show up.

---

3. The 1979 census reported that 2,312 people lived on Malo—probably a slightly larger number than when I was there.

Lessons Learned

- Don't be afraid to ask and follow up.
- Even blind optimism is helpful. Blind faith can work.
- Follow suggestions. It can make taking your next step easier.
- A woman alone can easily feel threatened, everywhere.
- Maternal advice is perspicacious; you should pay attention to it, but you don't have to follow it.
- Romantic love is not sufficient reason to marry.
- Take the opportunities that are offered to you—they could be critical for your success.
- Even the simplest tasks offer learning opportunities and insights.

Chapter Nine:

Plan B Is for Blood

A letter from Blumberg arrived soon after I settled in on Malo. His instructions stated that I "would not draw venous blood . . . [but could] in some cases draw blood by finger stick." He'd specified that I had to use "local health care personnel" to obtain blood from the vein. His attempts to control me and my research were an ongoing source of frustration.

This last-minute prohibition was annoying and ironic, since Blumberg had insisted I learn how to draw blood in the first place. That had been the plan. He'd arranged for me to go on morning rounds at Jeanes Hospital. I'd spent weeks being instructed by the phlebotomists as I drew blood from elderly bedridden cancer patients with collapsed and tiny veins. Then, once I was in the field, Blumberg pulled the rug out from under me. I summed up his memo as "You can't draw blood because you're not a medical doctor."

I was caught in a double bind, a research catch-22: I couldn't take blood, which meant that I wouldn't be successful, and if I drew venous blood, then I would be ruined for disobeying a direct order. It was a lose-lose situation. Blumberg's vision was that ICR would do the lab tests, which would require relatively large amounts of serum. Those amounts were possible only if I could fill the vacutainer tubes[1] with venous blood. In my trunk filled with laboratory equipment were thousands of vacutainers and needles. There was also the other equipment Scott had told me to buy. The Cuban WHO physicians offered to organize a team to visit Malo and take blood for me. Blumberg loved that idea—all those MDs—but the team never materialized.

1. A vacutainer is a test tube with a vacuum seal that pulls the blood into the test tube when a needle pierces the rubber top. It is a safe, convenient way to obtain relatively large quantities of blood.

Putting Theory into Practice

Blumberg's new requirements for drawing blood showed me where the power lay, but I also recognized when salvation came. Scott's equipment list included a plan B. The relationship I'd built with her out of curiosity and friendship rescued me.

When Scott arrived, I recounted Blumberg's restriction to her. Immediately, she said, "You don't need to use vacutainers. Just use the capillary tubes because the blood from a finger stick is enough." She described the basics, and the next day, during breakfast, she said, "It is time to put theory into practice. Let's start with your next-door neighbor."

I collected the equipment, and we walked the fifty yards to Jole's house.

"Jole," I said. "Can we take a blood sample from you and your family?" Vomule and five of their six children were having breakfast in the cooking hut. "Jole, can I start with you?" I sat down on the bench next to him. "This shouldn't hurt. It is just a finger stick. I'm going to put a drop of blood on this slide to test for malaria and then the rest of the blood will be used to test for hepatitis B and maybe some other substances."

I pulled out the slide, the capillary tube, and my book and labeled everything with a Sharpie.

"Hold his index finger and wipe it with the alcohol swab," Scott instructed. "Now angle the finger downward. Pump it a little to get the blood down into the fingertip."

I did as instructed.

"Now open the lancet, hold the fingertip up, and use your thumb to push the blood to the tip. Prick the fleshy part hard and remove the lancet. Let go of the finger, then get the capillary tube and milk the finger, collecting each drop of blood into the tube."

I plumped up the end of Jole's finger. Then I took the lancet and pricked the skin. Nothing happened. There was no blood.

"Try again," Scott said. "His skin is thick because he works in the garden. You have to use force."

"Sorry, Jole." I made another attempt with a new lancet. This time, I pushed hard. Jole winced.

"Let go of the finger," Scott said. "Wait a second . . . now milk it."

The first drop formed, and I placed it on a slide for WHO. Then I milked the thick, rough index finger and another red ball appeared before being sucked into the capillary tube. Five drops later, the capillary tube was filled. I put a plastic stopper on the end and placed it upright into a numbered plastic test tube with a screw top. The tube fit securely into a wire test-tube rack that I carried in my book bag.

When I stopped pumping Jole's finger, the blood stopped as well. I wiped it off with alcohol and offered him a bandage, which he declined. One by one, as Jole called out their names, each member of his family came and sat in front of me so I could collect their blood.

Next, we walked down the road to the second family, carrying enough equipment for forty samples. It became a repetitive process—stick, let go, bleed, squeeze, let go—milking the finger like a cow's udder. We used the fingers mostly, and toe or heel pricks on a baby.

Scott gave a running commentary as I practiced: "No, you're pushing too hard. Be gentle, gentle. Then the blood will flow."

The goal was to collect at least three drops of blood from each person, enough for test results.

After collecting twenty samples, we returned to my house and processed the blood.

Within hours, the blood clotted, with the heavier red cells settling to the bottom of the capillary tube, leaving the transparent yellow serum on top. Refrigeration did not affect this sedimentation. Next, each sample was removed from its test-tube holder. I used a file to score the tube at the red–yellow divide and then snapped it in half. One tube now contained only red cells and the other only serum. I emptied the red cells onto a numbered piece of filter paper and taped it into a composition book to take to ICR for ABO and G6PD[2] testing. I used the serum portion to test for hepatitis. Fortunately, my laboratory equipment provided reliable results using small amounts of both serum and reagents (0.025 milliliters of each). I performed a few simple tests to determine whether antigens and antibodies were present as well as the concentration (titer) of each positive result. I took Polaroid pictures to create a permanent record of the results.

It was a combination of science and practicality, a lesson in the beauty of simplicity. While Scott was there, we tested only for the antigen[3]

---

2. ABO is the blood group system described as "blood type" and could be used to verify biological parents and siblings; G6PD (glucose-6-phosphate dehydrogenase) deficiency can cause hemolytic anemia and, in some circumstances, can offer protection from malaria.

3. We used the Abbott Laboratories Auscell test. The anti-HBs test was not commercially available, but it was built on the same principles as the Auscell test for HBsAg.

because the antibody test had not yet arrived. Fortunately, the principle and method were identical. When possible, ICR tests confirmed my field test results using different reagents and assays. The good news was that the results agreed.

While we waited for the test results, Scott described how I could use similar principles to separate serum and red cells with a vacutainer: Turn the tube upside down so the stopper is on the bottom, allowing the clot attach to the stopper. Once the clotting is complete, gently turn the vacutainer upright, letting the clot dangle down. Carefully remove the stopper and clot. Let the few remaining red cells settle on the bottom of the tube. Pipette the serum and save. Simple—no need for centrifuges or electricity. Just gravity.

The next day, we got the results for the first batch. Remarkably, sample numbers one, two, three, four, and five were all antigen positive and chronic carriers. Scott and I stared at each other in disbelief.

"You're going to have a *lot* of positives here," Scott told me. "You're golden."

From that moment, I knew there was a high prevalence of HBV on the island. The islanders were asymptomatic carriers, which was partly why they were interesting. Despite the high prevalence of hepatitis B, there were no obvious symptoms in the population, which is characteristic of endemic disease. In the case of the Malo Islanders, Scott and I speculated that early and frequent exposure to the virus contributed to the high number of carriers.

"There are two different patterns of exposure to HBV with potentially different long-term outcomes," Scott said. "Children who become chronic carriers before the age of two seem less likely to develop primary hepatocellular carcinoma later in life. Kids who become carriers after two seem to be at higher risk of getting liver cancer later in life."

"Like in Taiwan," I added.

"Right. Taiwan has one of the strongest associations between HBV carriers and liver cancer. Your challenge is to find out which group the Malo Islanders belong to."

If one could prevent childhood infection with HBV, the incidence of liver cancer would theoretically be reduced, but it took decades for liver cancer to develop—it would be forty to fifty years before this benefit was realized. In 1975, the only treatment for hepatitis B was rest. There was no vaccine and no antiviral pharmaceuticals.

When I said goodbye to Scott at the airport, I was confident I would be able to do the serology, the data-based part of my dissertation. I knew what to do. The equipment was sterile, and the results came out in ways that were controlled for contamination—it all made sense. There were fail-safe parameters,[4] and I'd built positive and negative controls into the procedure. I took comfort from that. No one could accuse me of getting the wrong results, because of the system Scott had created and shared with me.

Blumberg had cleverly crafted the structure of my thesis so I would have a viable dissertation even if just one part of the project worked. If I only collected blood samples, then I could describe the epidemiology of HBV on Malo Island. If I only had information on parent–child interaction, then I'd have an ethnography on child-rearing. He saw it as a win-win. However, I didn't see it that way. I was not on Malo to do ethnography. It was critical for me to get blood samples. Blood samples were data—that was why I had opted to work with Blumberg.

Everyone agreed that my thesis would be groundbreaking if the epidemiology and ethnography came together. I would see whether my data supported the hypothesis of maternal transmission of HBV while generating culturally based observations that could explain both the epidemiology and the natural history of infection. That would be the trifecta—but it was a long shot. I don't think anyone expected it to happen.

Scott had provided me with the tools and the means to obtain and analyze blood. My next challenge was how to access people *for* their blood. Thompson offered a solution:

"Come to the dispensary and when people visit, you take their blood."

Islanders living within walking distance came to the dispensary for bandages, shots, and medicines. I showed up with my blood-collecting kit and finger-pricked whoever came to the dispensary. People trusted Thompson. He was a Malo native and part of the Avunatari social fabric. Among his duties was frequently traveling to the other villages to provide routine health care and update his records. For this, he had access to a truck and driver.

Thompson guided me. With him as my passport into the world of the island, my research on the natural history of HBV was underway.

---

4. Scott assured me that I would be fine if I wiped blood off myself with alcohol, and she told me I should cover my cuts or open flesh with Band-Aids. She turned out to be right. I followed her recommendations and never got infected.

FIGURE 29: Laboratory. The small hut between the kitchen and main house.

FIGURE 30: Laboratory setup. Test tubes (front) and Coleman lamp.

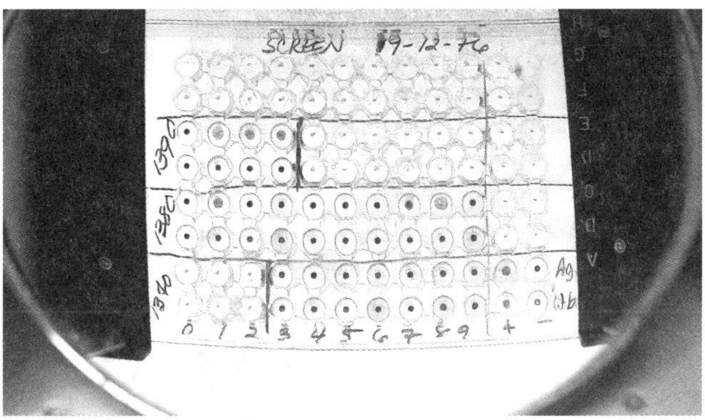

FIGURE 31: Laboratory test results. Hemagglutination readout for HBsAg and anti-HBs screening.

Lessons Learned

- There are always ways to circumvent edicts from people in authority.
- People are happy to help, so let them help you.
- Accepting help is a sign of strength, not weakness.
- Sometimes force is necessary to achieve the desired result.
- Simplicity is genius and beautiful.
- Good scientific methods withstand scrutiny even under primitive conditions.

Chapter Ten:

Drop by Drop

Both Penn and ICR had approved the proposed verbal consent informing islanders why I was taking their blood and how the samples would be used. Thompson or I would convey the information and obtain their consent. My priority was always using the first drop of blood for the malaria test and then collecting the rest in the capillary tube for hepatitis.

Children often complained, but their parents supported having their blood taken. There was a reasonable stream of people at the dispensary, so I was accumulating samples day by day.[1]

"Mark down everyone who is in a family in your book," Thompson told me.

I began by writing down the members of the family, both living and dead, along with birth order and, in the case of multiple marriages, which children were from which union, and the family members' approximate ages. I also recorded where people lived. I wanted a comprehensive catalog of people on the island so I would know from whom I still needed to collect blood and so I would have a system to show when I had obtained each sample. When I took a person's blood, I'd write a unique number next to their name in my green composition book along with their age, gender, and village. My results book listed numbers but no names. One secondary outcome of my research was collecting genealogy information from the adults. My system enabled genealogies to be reconstructed for the entire island population.

As time went by, when someone came in to see Thompson, I would know whether to say "I need you." Thompson and I could see where there were

1. When I returned to Malo forty-four years later, I met middle-aged people who reminded me that I had taken their blood when they were children.

gaps in my green book, so he helped me to fill in the missing pieces of the serology jigsaw.

"Let's go and collect some missing samples," he'd say. "Come with me and we'll get the family members you need for your book."

Off we went, usually at dinnertime so we could be sure they would all be home. The white canvas book bag slung over my shoulder was filled with the test-tube rack, lancets, and capillary collecting tubes; I was a vampire with a finger fetish. We'd go from house to house and bleed people on the spot so I could collect blood from entire families or find the family members I'd previously missed.

Walking Distance

My collection area expanded beyond Avunatari. First, we walked and visited families living in villages on the plateau (Avunarani and Avunaleleo). Next, we targeted the more distant coastal villages (Nanuhu and Avunambulu). Thompson, Vira, and I would go together, walking up to five miles each way in flip-flops.

Each time Thompson said, "I have to go to another village—come with me," I would go. He knew where everyone lived, and it became a systematic process. The island of Malo is thirty-four miles in circumference with thirteen villages (defined by the presence of a school or church, with most people living within a five- to forty-minute walk from the village center). Seven are traditional Tamambo villages. The other villages were founded by immigrants from the islands of Malekula (Tanmial, Avunarara, and Amambelao), Ambrym (Aviaboe and Asanasa), and Aoba (Nanjingo). I collected blood in all of them.

We'd ride around the island in the back of a flatbed Toyota pickup truck, occasionally passing the French plantation, where I bought a liter of fabulous Algerian rotgut (oops, red wine!) for next to nothing. The "wine" was pumped from a fifty-five-gallon drum into a repurposed glass jug plugged with a previously used cork stopper.

I ended up taking blood samples from over 85 percent of the island's 2,200 people and reached the point where I knew whose blood I hadn't yet taken. So, if I saw a person walking on the road or in their garden who had eluded my finger stick, I'd stop and, if I could, take their blood. I carried my white canvas kit everywhere. I quickly learned that if I didn't seize the opportunity

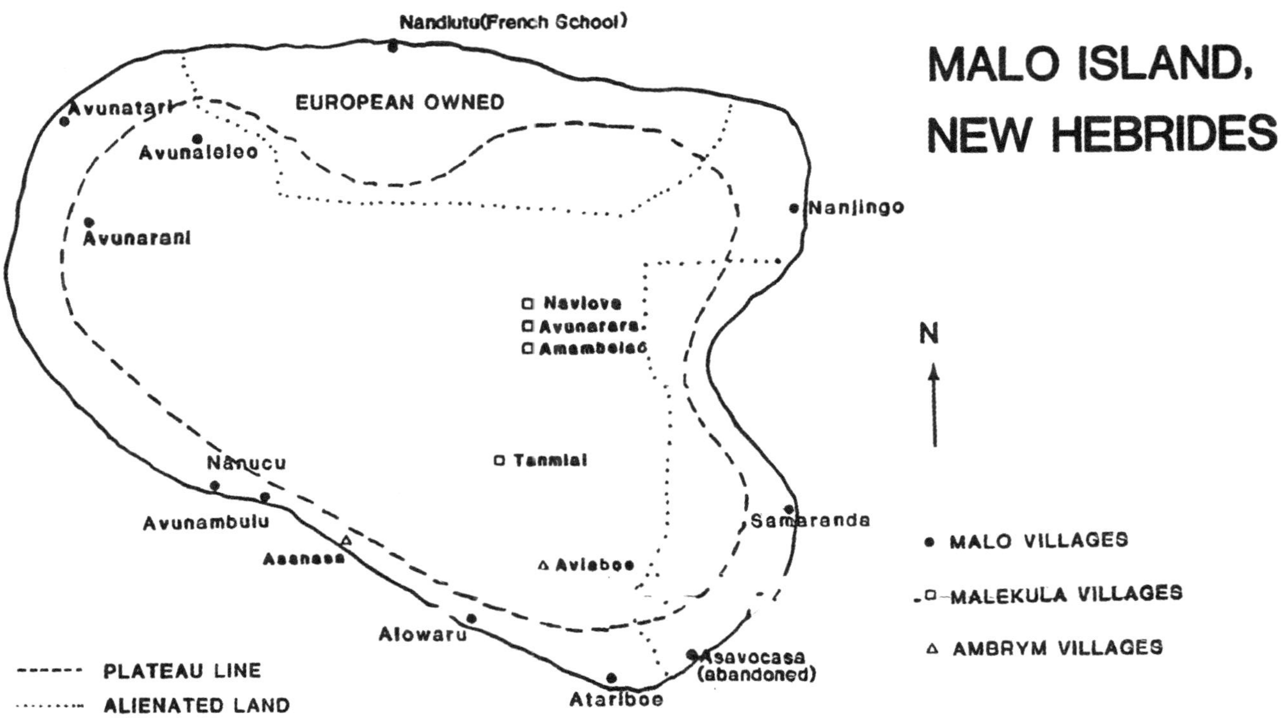

FIGURE 32: Village and ownership map. Native villages, immigrant villages, and plantation lands (1976).

when I saw someone, it took so much more time and energy to try to find that person again. There was a lesson in that. If you see something, grab it now—don't wait for a better time. Go out and get it while you can.

"It's just three drops!" I'd say, emptying a capillary tube I'd filled with water.

It became a dance. I always put the malaria test first and foremost. It was a disease they knew and understood and for which they could be treated. Once the finger was pricked and a drop of blood had been put on the malaria slide, the remaining drops were collected for hepatitis testing.

Sometimes people asked whether I was selling their blood.

"Three drops!" I would reply. "How could I sell that?" Older men, mostly those who were part of the Nagriamel political party, would say I was selling the blood and that I was an agent of the British government. When that happened, I presented myself as a student working on my own. I also reminded them about the benefits of consenting. Generally, they agreed to let me take their blood.

A few months after my arrival, a cultural anthropologist—Bob—arrived and built a hut next to my house. For twelve months, we shared the kitchen.

FIGURE 33: Bob (right) under a tree on Malo

He was a graduate student from Bryn Mawr College doing his fieldwork on the traditional customs of the ni-Tamambo. Fortunately, he was able to write the ethnography that I couldn't and wouldn't.

Co-op . . . eration

On Malo, alcohol consumption would occur either in a house or outside the cooperative store where men drank beer and played cards. There were no keggers or cocktail parties. Alcohol was too expensive to share. Santo had a wide range of alcohol available for purchase while the Co-op offered only beer and other imported goods. These "delicacies" included canned mackerel, fresh-baked white bread (from the Chinese store in Santo), shorts (also from China and only for the men), kerosene, flip-flops, machetes, bags of rice, matches, string, blankets, bottled scents, and batteries for shortwave radios. It was pretty much always the same stuff. One person would motor over to Santo in a V-shaped hulled aluminum boat, shop, and bring the goods back to Malo for resale.

The Co-op also was a savings bank. Since Malo wasn't a cash-based economy, savings didn't have much appeal. Out of curiosity, I opened an account and received a paper savings passbook showing my deposit of five Australian dollars (which I have yet to withdraw). When an islander had cash, it came from selling burlap sacks filled with copra or cocoa (*kakau*) in Santo or through the Co-op. Another way to accumulate wealth was to work in the nickel mines of Nouméa. Young men would stay there for a couple of years and return with enough money to buy a truck or a boat. Occasionally, someone like me would come along waving their cash, but no one could rely on that—there was very little money passing through people's hands.

The Co-op was the center of many activities, but it was tiny, maybe eight feet by eight feet. You didn't browse. You didn't need a cart to go up and down the aisle. There were no refrigerated drinks or sandwiches. If you were lucky, you could get two different types of mackerel: with or without tomato sauce. Buying the tinned fish, bones and all, was an easy option, I guess, but bizarre, since fresh fish was abundant, readily scooped or hooked from the sea. Then again, fishing required dragging a dug-out canoe off the coral, paddling out beyond the reef, dropping a line or net, waiting, and expending energy versus opening a can. I know which I would choose. Do you?

Lessons Learned

- Some US health care rules and procedures apply even in non-US countries.
- Nobody enjoys giving blood.
- Having a tangible, actionable reason for obtaining blood is important.
- Grab every possible opportunity to fill a gap in information or blood collection.
- Men enjoy socializing with alcohol near the place where it's sold.

Chapter Eleven:

Writing—Four Ways

In each of my letters, I let a particular side of my personality show. Communications with my professors were professional and unemotional. With Michael, I could be open and vulnerable, while my family members received superficial narratives. Letters to my friends and classmates begged for gossip while I tried to amuse them with local anecdotes. It was as if I was trying to sort out who I was on the page while trying to engage the recipient.

To Friends and Classmates

Letters to my friends and fellow students pointed out things I thought were cool. I emphasized my witty, interesting side and hoped a return letter would contain juicy gossip.

Tuesday, October 1, 1975
Dear Mary Louise,

Greetings from the land of the still noble but sick native. Virtually the entire island is sick with the flu, of all things! Bob had the winning mosquito in this week's dengue fever lottery, and he has been bedridden for about a week now. It really is a miserable disease. He keeps trying to impress on me that he isn't sick, but when he has to have help to make it from the bathroom to his house, he is sick. Poor person, he looks miserable.

Last week was great, virtually no work to do. Thompson hurt his leg walking to Avunambulu and Jole is sick with some kind of flu, so I have a holiday! I got a chance to test the blood I'd taken the week before and mostly I slept. I hadn't realized how close to exhaustion's door I really was. Now that all of the tests are here, all I do is work, and usually until well after midnight, but last week was terrific. I even got a chance to read. This week looks like a real workweek "back again" (a good pidgin phrase), and it is just as well. I can't wait until I hit a thousand samples tested.

Are you ready for this? I've been so busy that I haven't even had time to transfer entries to the data book and enter age and gender so I can xerox the results and send them to BSB and just write him and say that the results seem to be interesting . . .

I'm trying to change because I'm just so damn tired of fighting with him all the time. I hope this works. I wrote him a very long letter trying to explain what it was like out here and why I have been generally so nasty to him. He wrote back immediately a very nice and very humble letter. Maybe there is hope . . .

I also used these letters to vent my frustrations with my professors. Connie and I exchanged letters using our proper given names as a symbol of familiarity. She was a confidant of Blumberg's.

August 19, 1975
Dear Caroline,

It was absolutely great to get your letter. It is going to be difficult for me to respond, but I'll give it the old college try. First of all, there is no question that you understand BSB and that your analysis of him, socially, is correct and perceptive.

One thing you must remember [is that] with him, it is the explicit level that counts. I think that you are right when you say that he does care, or at least as much as he is capable of caring. I do not trust the man, and he is not above using you. One of his great problems is that his self-worth comes from how others view him, and right now you are filling an emotional void in that area. Also remember that his scientific work isn't faring too well in the eyes of many of his peers (don't let this get back to him, but it is the case), and many remain unimpressed by what was a real discovery and contribution.

From the other end, having to relate to him as a "mentor," I have recovered from my first lonely days and have become my jaded self again. I really cannot stand him. He has some nerve bitching about the cost of the ice box I need to store my samples. In fact, when I think about everything he either did or does to try to hinder the project, it makes me angry.

Cute things like threatening to cut my salary when he doesn't have to. He sent me out here unprepared to do what I am doing, and doing well, because he wanted to have total control over the blood data. Fortunately, Scott came and she ironed out the bugs. The project is going to fly like he never imagined. In fact, it may be the best corpus of data that he will ever get his hands on.

The main thing that taking the blood has done for me is to increase my self-confidence and my feelings of self-worth. The primary justification I have for drawing blood is that I'm taking malaria smears and collecting whole blood for G6PD tests, and I'm doing that for the natives. Because he has never done fieldwork, I don't think he is capable of understanding a lot of the things that I am doing. I feel very good about it. It is emotionally draining, but I am definitely growing and accepting responsibility, and the work I'm doing is distinctly medical.

I know that it is going to be a good dissertation, and [I know] that the data are solid and it'll be good. The other thing is, the thought of him coming here is horrifying. I know that he'll try to tell me what I'm doing wrong, if for no other reason than to make a "contribution" as [he believes] I have to be doing something wrong. I know there are things that could be improved, but he rarely picks those to correct. (E.g., among the things I ordered to take to the field were alcohol swabs, and when I showed him the list, ex post facto, he said to take cotton balls instead because alcohol swabs would dry out. The entire exchange was an ego trip, as he couldn't believe that I had gotten everything correct. So here I am with these damn cotton balls that are more trouble than they're worth. Then he has the nerve to write and ask when he should send the alcohol swabs, which don't dry out. It drives me up a friggin' wall.)

On the one hand, I am grateful that he has made it possible for me to do all this, because it is going to be so good. On the other hand, every time I think about him, it is with a great deal of distaste. I don't know if I'll ever change.

Pissed? Better believe it. The other thing is the conditions in the field (they're good, but it is a type of camping out). Scott pointed this out, things that I don't even notice, like the four-inch spiders that live in the toilet and around the kitchen, or the noise from the cows and chickens, and dogs, etc.

He'd go crazy. He's just not a field person and he'd think that the entire thing was just awful. I really am sorry to subject you to this outburst. Enough of BSB. I can't take it.

To My Professors

The letters to Blumberg contained none of the drama and emotion that filled my letters to friends; instead, they were professional and work-related. They also showed the struggles and reconciliations that I navigated as part of the blood and data collection processes.

Tuesday, September 9, 1975
Dear Dr. Blumberg,

I thought I would send you a quick status report. Since Scott left, things have been progressing very well, as far as they go. The most frustrating thing is that the antibody test has yet to arrive. Is it possible for shipping to have Pan Am put a search on it? It is terribly frustrating to just collect, slave over the sera, and freeze it. As of last Monday, I have some 350 samples, and so far, all the cooperation in the world here but no cells. It is driving me crazy.

About the problem of the vacutainers. There are a number of issues at stake:

- *I am the only person I can rely on to be available to take blood, and it will take months to do the entire population. Since I am limited by our agreement, I cannot go off into the wilderness needle in hand.*
- *People are very used to finger sticks here and not used to vacutainers.*
- *Vacutainers are only good for the adult population and about 60 percent of the population is under fifteen years of age.*

I would imagine that it will be very feasible to get vacutainers on all the positive adults. Especially as, when you come here at the end of November, you can bleed those people who are positive. The other option is to have the technician from Aoba or Aore come over here and take those samples. However, I don't see us doing this for quite a while.

The other frustrating thing is that without the micro-blood collecting tubes, I can very quickly grind to a standstill. It is not that I don't like to wait, it's that it is so frustrating!

That's about all for now, please thank Ann[1] for me, for taking care of everything while you and Tom were on vacation. I hope that Cape Cod was relaxing and that the new semester goes well at Penn. My best to your family.

Sincerely yours,

Betsy

To Michael

My letters to Michael were honest, heartfelt, and genuine. They also exposed my vulnerabilities.

Wednesday, July 22, 1975

Dear Michael,

I just got parts two, three, and four of the four-part series of letters. I can't tell you how much I enjoy and look forward to hearing from you. It sounds really silly, but I calculate about how many days it will be before I get a letter from you, then not so patiently wait out the days. (Am I not supposed to say that?) Every time, I have found it so amazing that you have dealt with what I have been thinking. It is rare that I receive such thoughtful and tailored responses to my concerns. You continually amaze me. For a change, let me respond to some of the things that you wrote.

The parts of your letter that you said that I might find uncomfortable were of course the most comforting. Funny how an odd word or phrase here and there can bring on an almost giddy response and produce a tingle of excitement. I mean, it is really neat, even though I feel pretty silly sitting and/or lying in bed with my flashlight poring over every word time after time, each time getting a thrill. And then, when it gets so late that I should sleep, I find I'm too wound up to even consider sleep. My mind "dances and leaps" into and out of confusion, fantasy, and reality.

It was funny, this morning (I got your letters last night) Bob (will

1. . Blumberg's administrative assistant.

elaborate later) and I had a very long talk about various things back in the States. He talked about his girlfriend with whom he has had a very unsatisfactory and tension-ridden relationship on and off for the last year or so. Then, well, actually his conversation followed mine, I tried to talk/ explain to him about you. The more I tried to make myself coherent, the stranger the entire thing sounded. It is funny, because there are so many different aspects to our relationship: age, school, family history, career. I know that Penn was the wrong time and the wrong place with me in grad school and you in undergrad. It was really strange too, Bob and I got to talking about how relationships would be different when we returned to the States. I got to do my own brand of philosophizing and started comparing the different attitudes that one has toward friends and lovers. I am convinced that the only really sound and reasonable relationships that have viability are those that are initially based on liking another person. Without that base, there really is very little. It is also interesting that when removed from [an] environment, the things that one remembers and thinks about are very revealing. This is going to sound really odd, but I am very glad that we did not become lovers, because right now that might shade and cloud the fundamentals of our relationship. It is really a luxury to be able to sync with a clear mind and not have to immediately bring in a physical relationship. Oh boy, I better stop. You know I'll be telling you some of the really inane things I decided years ago; e.g., if I had got remarried, I would not have slept with that man until we were married—ridiculous, but there is a real reason and a lot of emotion behind it.

Am I making any sense at all? I am so happy for you and the self-gratification that your move into the graduate program at the University of Toronto has provided. Those feelings I can understand very well and empathize with. There are times when I forget how vulnerable you are, too, for you always seem to be so sure of yourself. In fact, I have always admired the fact that you didn't follow the obvious course of action: go to college, then on to architecture, etc. I have always been the shining example of what I was supposed to be, not that it was ever articulated, and in my own way I rebelled a little, like by not graduating with honors and by just barely getting into schools, etc. Even though you feel "behind," I don't think it is important in the long run. Ironically, looked at from one perspective, a person is always "behind" their elders. This chronological fact can be a rationalization for many things, but I think things even out at around thirty years of age, such that being "behind" becomes a very relative thing and encompasses many different areas and emotions and actions.

I think I am going through another phase of coming to grips with life here. First, let me explain who and what Bob is. He arrived a few weeks ago as a grad student in anthropology at Bryn Mawr. He will be my neighbor for the next three months (until he finds a more remote place to live), and undoubtedly, he will become my friend and confidant over the next fifteen months. It is nice to have someone to talk to, particularly on a completely platonic level. When he first got here, he drove me up the wall and then we have gradually been talking things out. He's gotten me to realize some of the pitifully obvious mistakes I have made here. Funny how one can't see the forest for the trees sometimes. I am now beginning my disassociation with the white expat community in Santo, who were my lifeline when I arrived. Funny how in one sense I never thought of them as "colonials" but only as people to be evaluated as individuals and not in the generic. I think that I am right in this approach, but Bob has a real aversion to white persons. I am agreeable to going along with this and to trying to go the whole "native" route and try to rectify any mistakes I've made along the way by being involved with the Santo expats. I'm not sure about "going native," but I will distance myself more from the white community. I feel better about a lot of these things since I've come to this conclusion. And in many ways, Bob's being here has been very helpful, and I think I've learned more from him than vice versa. It is good. So much for "this week's insight." It is true, though, each week seems to offer some new insight and I can't help but think that I will be a better and fuller person for it.

I really enjoyed your comments on veni, vidi, vici. I had never even thought of defeat as the reciprocal of conquer. Fantastic. I think that I am gradually getting my act together and that an absolute defeat or victory is not important but that uncomfortable mediation is.

Well it is already 1:30 p.m. and I haven't done anything all day. As I won't be going into Santo until Friday, I'll probably continue this later on. Take care for now.

To My Family

Finally and importantly, I wrote letters to my family. Those letters reveal the foibles of a daughter sharing emotions, medical issues, and observations on Malo life. I received advice from my mother, and most of my correspondence was with her. However, I also corresponded with my younger brothers, with whom I shared my wisdom. They reciprocated with updates from home and the occasional hand-drawn puzzle for my amusement.

Saturday, June 14, 1975, Queen's Birthday
Dear Darryl,[2]

A very happy, if somewhat belated, birthday. You're fourteen now and almost through junior high school. How the time flies. It really does seem like yesterday when I was just going into ninth grade and you were born. We were still living in Parkmerced and were thinking about building on Twin Peaks. It certainly doesn't seem like fourteen years ago.

I remember when I was fourteen, or had just turned it. We moved into the house on Mercedes Way and I had to begin [coping] with making my way to school without all of my friends and buddies from grammar school. It was traumatic at first to have to walk those eight lonely blocks to school and miss the frivolous and carefree conversation and gossip on the Parkmerced school bus. But obviously I survived.

I also remember those years, and I don't know if it was eighth grade or ninth, as being very lonely. I think then I was in my hate stage. At that point in time, I hated Mom, I hated Sal, and I don't think that I was any too keen on any of my friends, except Dawn, who at that time was close to Candy and not to me. Although I never spent a lot of time alone then, I felt very lonely and friendless, as sort of a fifth wheel. But fortunately, all that passed. Nonetheless, I remember those years with a certain amount of pain.

I think the nicest thing about being fourteen is that you're just that much closer to getting through the teens. It is a time of a lot of changes, both mentally and physically. It is a time when you're particularly susceptible to ridicule for things that are a natural part of growing. There are times when all you want to do is hide and come back seven years later as

2. My brother, who is twelve years younger than I am.

an adult. But unfortunately, part of the growing process involves learning to handle these situations and to face them squarely.

Another good thing about fourteen: It was nowhere near as bad as being thirteen. By fourteen, I had given up all pretense of being a horse but not the desire of owning one.[3] *I had not yet discovered boys and wouldn't for a few years. What did occupy my time was swimming, and those times were fun. I guess, when I look back on it, for me fourteen was coming out of the hating days. In a sense it was like a rebirth and I began to like people again, and myself even just a wee bit. No, fourteen was good overall, but it was hard, and I wouldn't want to relive it.*

Most fondly, fourteen was discovering makeup and convincing yourself that you didn't look a childish fourteen, but in fact presented the world with a sophisticated, worldly-wise sixteen-year-old. The delusions were grand and looking back, quite funny, but it was the luxury of living them that was good. Well, again, Happy Birthday. I miss you and I love you.

These four Betsys drew upon different facets of my personality and sustained me through the roughest part of my fieldwork. I faced many challenges: isolation from home and my support systems, learning to navigate the diverse cultures of the expat communities in Santo and on Malo Island, learning how to do field-based epidemiology research, and dealing with logistics and supply problems related to essential equipment.

Letters were vital and made fieldwork possible. They also raised questions about what would happen when I returned back to the States for rest and relaxation over Christmas.

3. Dawn and I would pretend to be horses on our hands and knees. We would take turns pretending to ride one another. My mother described me as "half of two horses," and when my fifth-grade teacher met my mother, he noted that I was one of the "horses." See "Graduation Antics," posted July 12, 2009, by malovanuatu, YouTube, https://www.youtube.com/watch?v=bzv5HxcD0rw.

Lessons Learned

- Have the courage to vent—it is cathartic and helps put issues into focus.
- It is a gift when someone is honest and direct with you—it provides the information you need for next steps.
- People in authority can create no-win situations without consequence for themselves.
- Conflict forces solutions.
- Letters allow you to foster vulnerability and communicate much more than you ever would in person.

Chapter Twelve:

Alone and Adrift—Casual Sex Is a Bad Buoy

Santo provided a definite contrast to Malo—it was considered a hotbed of city life, a place where I could do whatever I pleased. As one of two single women in Santo, a town of four hundred Europeans, that's what I did. I'd go out, get drunk, and sleep with the bank manager or end up with a guy from the Seychelles. The Australian cowboy said I was completely out of place and should go back to America, and I screamed, "Fuck you! I *am not* out of place!"

But I was.

I was caught between two worlds; in some ways, my brain was still back in the US. The expat world was more confusing than was the islanders' world. I thought that because they could speak English and were white, the expats were like Americans. But they were *not* like Americans. A Brit is not an American, neither is an Aussie, neither are the French, and neither are the New Caledonians. Australians speak the same language, but the meanings, norms, and expectations are completely different. When an Aussie got "pissed," I thought they were mad or angry about something. Later, I learned it means being drunk. I rationalized that it was because pissing is what happens after too many beers have been *drunk*.

If the expats did something I thought was nice, I assumed they liked me. More often than not, I'd misread the cues and they didn't really like me. The worst was when a guy became aggressive toward me because I'd misread the signals. This happened with the Australian cowboy.

"Why are you here?" he asked. "I didn't ask you to meet me at the airport."

"I wanted to surprise you. So I borrowed the bank manager's car to pick you up."

"That's too controlling!" he said angrily.

"I thought that was a nice thing to do," I replied.

"It was not a nice thing to do—you are being controlling."

"I'm sorry, I had no idea."

I did end up giving him a ride, but that relationship was doomed. Later, I had a months-long relationship with a widowed Australian who was good fun but was also a person who suffered from alcoholism with three children to raise.

The French were less of an issue because I didn't speak French. When they said, "I love you," they meant "I like you," which was a problem—at first, I thought the next step would be a marriage proposal.

With the New Hebrideans, there was little room for assumptions. If they said something was *bugger-up finish*, I had to ask what that meant (*all same what 'em*). I knew *finish* meant it was over, but I didn't know if *bugger-up* was the pidgin equivalent of "bugger all" (British slang meaning "nothing") or if it was a reference to anal intercourse or what. Probing and context often clarified that *bugger-up* meant something had stopped working. Since we had no common mother tongue and our cultural backgrounds were very different, I paid attention to everything. It was also mentally draining. When everyone speaks using their second language, like Bislama, it is easier not to assume what is being communicated. With people for whom English was their first language, I made too many assumptions and soon discovered that as an American, I was separated from the Brits and Aussies by a common language.

There were other notable differences between the two colonial powers in New Hebrides. The British were standoffish. Many of them had been raised in Africa and had made their careers in the Colonial Service. These civil servants were colloquially described as "Rhodesian retreads," referring to whether they'd served in what is now Zimbabwe. In general, they made a point of never speaking any of the local languages. The exception was Dick Baker, the BDA for the Northern District.[1] If (and when) Brits liaised with a

1. Francis Eustace (Dick) Baker was Oxford educated, and prior to his time in New Hebrides as the BDA of the Northern District, he served in the British Solomon Islands Protectorate. Later, he was chief secretary to the Falkland Islands government during the war with Argentina.

local woman, they were discreet. The French, by contrast, all spoke Bislama, and many even learned the local languages. They freely associated with the locals—one result was an abundance of light-skinned Melanesians. The derivative Australians and New Caledonians tended to follow the French model.

When I described these Anglo–Franco differences to Goodenough, he thought they had prehistoric origins. Goodenough had also studied European prehistory and noted that France had been an agrarian society. He speculated that the French were culturally invested *in* the land and *in* its people. On the other hand, the prehistoric Britons had been more of a pastoralist society and were used to taking *from* the land. It was an interesting point.

Regardless, I found that the French were easier to relate to than the British.

There, but Not There

In the beginning, I left the island once a week for Santo, where I acted out my insecurities, and that made it harder to integrate into Malo culture. I wrote so many letters that I paid someone to drive me to Santo in their boat to mail them. My stipend made me wealthy, so I could afford the trip. My money bought me services, and I had not yet learned that respect is earned, not bought. I would usually be on the boat by myself, but I sometimes traveled with one or more freeloaders. I was used to the crossing, and, even when it was bumpy, I no longer worried. I could swim, so I wasn't afraid, and, importantly, I'd decided the ocean was safe, since I'd never encountered a sea snake when I swam.

On the island, I was more the vulnerable Betsy, as I didn't know my way around. All I wanted was to have the language skills of a two-year-old. That would have helped so much. I told myself that if I stayed for two years, I could reach that level of competence. That comforted me. I rationalized that I was every bit as smart as a two-year-old. I also smiled a lot because I felt so vulnerable and foreign. I wanted the islanders to like me, but without knowing Tamambo, it was difficult. I dressed differently, looked different, and spoke only Bislama.

I wasn't integrated, but I was accepted. Everything seemed so difficult at times. I sat and watched people meet and interact in Tamambo. There was nothing else I could do. I was on the periphery with a smile fixed on my face. Only a handful of the islanders spoke English, but most often, we spoke pidgin English. It wasn't the same as knowing the local language.

On the island, I wouldn't participate per se. I would observe. I'd be there physically, but not there mentally, because I was so preoccupied with thinking about how wonderful it would be when I got back to America. All I wanted to do was think about what would happen when I returned to America and saw Marty. Or what would happen when I visited Michael in Toronto. I'd bared my soul to him, and he was my most intimate correspondent. I thought—we both thought—that maybe, through our letters, we could develop something more. My head was completely back in America for those first seven months while my body was in New Hebrides. Even though I read books and did things on Malo, they didn't really register. I was killing time.

Every week, I made sure I collected and processed at least one hundred samples. Setting and meeting a goal kept me focused. It took three to four days to collect the samples and a full day to process and test them. At the end of the workweek, I felt I had accomplished something, had made progress in my research, and deserved a reward—having a good time in Santo during the weekend.

I had my dalliances. I got drunk and went with the flow with those men who caught my eye, and it seemed innocuous. I would go back to the island on Sunday evening or Monday morning and change back into Island Betsy, leaving Party Betsy for another workweek.

My first wake-up call came that fall after I got infected with dengue. The first night, I had a high-grade fever and my dreams were incredibly vivid; something on Malo was coming toward me or trying to hurt me. It was a warning.

When I recovered, I went to visit the French teacher and his wife on the other side of the island. I walked to their place and spent the afternoon waterskiing. All I could think was, *Waterskiing is fun! Partying is great! Let's just have a good time.* People bought me dinner, and I drank champagne. That night, I saw that as the other Betsy, I was on the island doing something five days a week, but for two days a week I was with Europeans, which felt more real, more familiar, more *me*. I was anything but engaged in the local culture.

After dinner, the French teacher invited me to walk outside with him along the beach. "I wanted to talk with you about the reputation you are getting," he said.

"Reputation? What kind of a reputation?" I asked.

"Your activities in Santo are not good. I know you are here to do serious work, but other people think you are only here to party."

"What would you suggest I do?"

"You could spend less time in Santo or be more discreet."

Oh, this is bad, I thought. *It's really bad. Time for a change.*

Spillover

I slowed the pace of my Santo-based activities. But then, less than a month later after my conversation with Bob about "going native," the impact of my transgressions became serious. One of the natives came up to me and said, "Betsy, you're in trouble, and you better figure out what you are doing fast. You really shouldn't be doing that. You shouldn't be up at Malo Pass."

He directed his gaze toward the other side of the island, where the French teacher lived, implying that I was having sex with someone there. I wasn't, but it reinforced how close to the edge I was, how everything was falling apart. I was being accused of things I wasn't doing, and it was all converging. It didn't matter that I hadn't had sex with the French teacher; the perception was already there.

Was I angry at that almost Victorian notion of a woman's reputation? No. That didn't bother me. My career was on the line, and *that* bothered me a lot. I thought it was odd that my Santo behavior had a spillover effect onto my professional side, and that was very, very threatening and very, very scary.

He went on to say, "The British think you are only interested in Black men. The French say you don't like men and prefer women. There is lots of talk."

I was too stunned to defend myself. "Thank you" was all I could say. I walked back to my house and thought his last words were more of a cultural statement about French men and English men.

In retrospect, it was interesting because I was caught in the public/private dichotomy that is integral to anthropological studies. I was in the vortex of a microcosm of my own making. I had a choice to make, and I did. Santo Party Betsy became Santo Serious Betsy. When my party persona stopped, the gossip stopped. Soon thereafter, I finished up the bloodwork.

Worlds Apart

No one talks about fieldwork; it's rare to find honest descriptions of what people really go through. They will tell you what they discovered but not what they personally experienced. Yet fieldwork is so very personal. You endure so many changes, and you make so many mistakes. It is very easy to sit in a classroom and be told you are going to become a participant observer, you will become part of the culture, and you will have a wonderful, life-changing experience. No one mentions how difficult the experience will be because you are clueless or the frustration you'll feel when you don't understand what is happening. You arrive as an outsider, ignorant, a child, an authority figure—a confusing bundle of attributes. You're a zillion different things, and none of them bear any relationship to who you are, or were, as a student in your own culture.

As both a scholar and a white woman with money, I was different from everyone in Malo's subsistence society. We were many worlds apart, but I hadn't been briefed on the adjustments I would need to make. Islanders would ask how many siblings I had, whether I was married, if I had children, if my parents were alive—everything was focused on family and relationships. That was their frame of reference. They didn't have the words or the concepts to understand my other world (academia) or ask about my research, and I didn't have the words to accurately describe it in pidgin English. I would say there was something inside of blood that could make a person sick. They understood malaria, but hepatitis was not a disease they knew, so it was difficult to describe. They trusted me when I asked for consent and took their blood.

I had no language skills, but I was making vocabulary lists and taking language lessons from Jole. Bob was making progress on the language and talking to people who were still involved in *kastom* and living traditionally. However, my relationship with Bob was gin rummy–based. He was my opposite. He engaged in the culture, rarely left the island, and was present in what he did.

I went to some local events. Vira, who worked for me, would say, "Let's go to a *bong sangavulu*." This happens on the tenth day after a person's death. It recognizes the end of the mourning period. It is a feast for the bereaved family and signals the return to normal life.

I thought it was a good idea to go to big events, especially if they were within easy walking distance. I watched the ceremonies that combined both traditional and Western religious elements and saw the color of it all. It was interesting, but these activities were not relevant to my research. The ritual pig killing was always dramatic, as was the casual way the pig's bladder was removed and given to the boys so they could use it as a soccer ball.

But what struck me most was the separation of the genders. Women and men were on separate sides of every event, even at family meals. It reminded me of high school dances with the boys and girls standing on opposite sides of the gym or cocktail parties where same-sex groups naturally congregate in a circle around the appetizers or the bar. In each of these circumstances—dances, cocktail parties, events back home—I usually made a point of inserting myself into the men's group because I found it more interesting. Gender-based separation was a cultural artifact that was easy for a Westerner to understand. In retrospect, I think it was something I thought I understood but actually didn't—like being separated by a "common" language, I projected a common meaning to a "common" behavior.

There were other aspects of island life that even today could seem shocking to outsiders—such as the married fourteen-year-old mother with her two children. That's just the way it was, and even in parts of the US, it wasn't uncommon one or two generations ago. Often, what initially seems strange in a foreign culture isn't odd once one objectively examines one's own culture. Frequently, there are similarities.

There were imposed constraints that kept me on the sidelines for some conversations and activities. One condition of my residency permit from the British government was my agreeing, in writing, that I would neither engage in politics nor express any political opinions. Notably, the country's first elections were held that November. It was a big deal. The men dressed up to vote. Thomas, one of the older men in the village, came to the voting station wearing his most treasured clothing from the war: a khaki shirt and a gas mask. After voting, he took them off. The symbolism and respect the islanders showed for the event made me proud to be American. Politics was a hot topic of discussion in 1975, but there was no mention of the country becoming independent until the following year.

Toward the end of my first year in the New Hebrides, I started to have reasonable friendships in Santo. I became good friends with a French woman psychologist—Meythe—and an Australian—Heather—who was a nurse with five young children. Her husband, Ted, was an accountant for the government. These were people I could talk to, people with whom I could be myself: a Betsy who was a combination of everything rather than a performer accentuating just one part of my life. Those women had stabilizing effects on my behavior, too. I was starting to see that the women in my life made me who I was.

After seven or eight months on Malo, my relationships started to go deeper than just grabbing a person, taking their blood, and testing the blood.

Meanwhile, I was having so many conflicts with Blumberg. He wanted me to end my fieldwork in December, saying I wasn't doing what was expected of me. I knew that meant I wasn't doing what *he* wanted me to. He'd always thought one year of fieldwork was enough, which it clearly wasn't.

"That is just not true," I wrote to him when he said I hadn't accomplished enough. "I'm doing finger sticks, I'm getting the samples."

"That doesn't count," he replied. "I'm withdrawing your funding."

And he did. It motivated me to write my own grant application to the National Institute for Mental Health, which was offering predoctoral support for anthropology, while I was in the field. It forced me to analyze the data I had already collected. It also taught me that you have to stand up for what you know is right, and I *knew* that Scott had put me on the right path no matter what Blumberg thought. But I would still have challenges to deal with when I returned to Penn after Christmas.

Lessons Learned

- Bad behavior catches up with you.
- Changing your behavior changes people's perceptions of you.
- A common language does not mean a common culture.
- Having a different language and heritage from someone results in making fewer false assumptions about that person's culture.
- What foreigners think of you reveals more about them than about you.
- Money can buy services but not respect.
- Listen to people who have the courage to offer criticism.
- Women friends are a source of stability.
- Stand up for what you know is right for you.

Chapter Thirteen:

Christmas Chasm

By the time I returned to the US for a break in December 1975, I'd had some success with my data, using, believe it or not, punch cards! These cards had numbered holes around the periphery, and the presence or absence of a trait was indicated by punching the hole open (presence) or leaving it untouched (absence). It was a very early form of manual data analysis of binary information, the type of cards used in early computers.

I created a code, punched my McBee KeySort cards with the epidemiology data, and did a preliminary sort using the long needle that accompanied the cards. The antigen-positive cases would fall out and the negatives would remain on the needle. It was simple and easy. This was my first epidemiologic analysis of the serology.

I discovered that the prevalence of the hepatitis antigen varied among the ni-Tamambo. The villages with the highest prevalence, 24 percent, were in the north. The percentage decreased as one traveled down the western coast to a low of 10 percent in the south, with a low of 4 percent in the eastern villages. The immigrants from Ambrym (16 percent) and Malekula (26 percent) also had a high prevalence of HBsAg.[1] At the time, the northern ni-Tamambo and Malekulans may had the highest prevalence in the world.

---

1. See Appendix I for a map of the villages on Malo. The prevalence of the antigen was as follows for the ni-Tamambo: northern villages (Avunatari, Avunarani, Avunaleleo), 23.7 percent; middle villages (Nanuhu, Avunambulu), 20.4 percent; southern villages (Alowaru, Atariboe), 10.5 percent; and eastern villages (Samaranda, Nanjingo), 4.3 percent. The Malekula villages (Tanmial, Naviova, Avunarara, Amambelao), 26.5 percent, were located on the eastern/inland side of the island; and the Ambrym villages (Asanasa, Aviaboe), 16.4 percent, were located in the southwestern/inland portion of Malo.

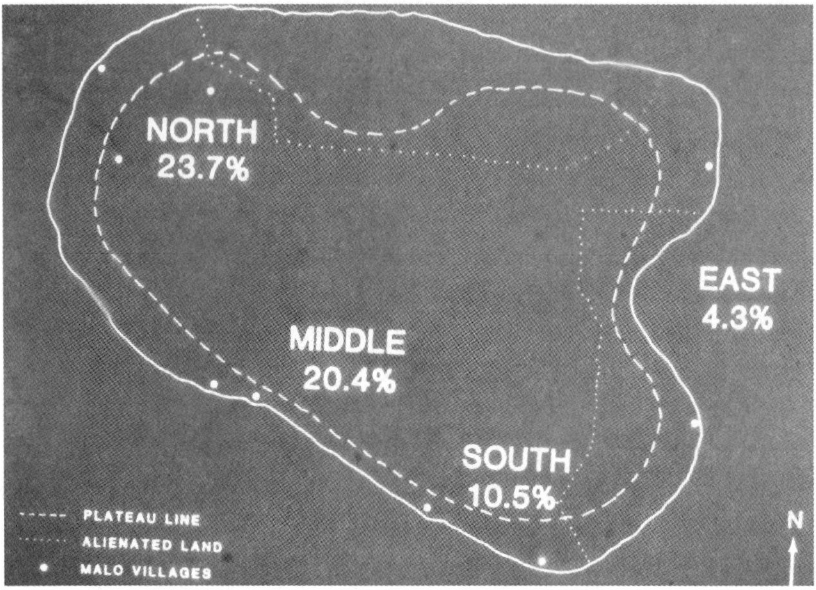

FIGURE 34: HBsAg frequency among the ni-Tamambo by geography

Distribution of HBV characteristics for each ethnic group living on Malo Island, New Hebrides. Information on the total resident population and the percentage sampled for each group is also included in brackets.

	Ambrymese	Tamambo	Malekulans	Total
HBsAg(+)	10 [16.4%]	209 [16.4%]	44 [26.5%]	263 [7.5%]
anti-HBs(+)	9 [14.8%]	227 [17.8%]	33 [19.9%]	269 [17.9%]
sero-negative	42 [68.9%]	839 [65.8%]	89 [53.6%]	970 [64.6%]
Total sampled	61	1275	166	1502
Total present	67	1537	363	1967
% sampled	91.0%	83.0%	45.7%	76.4%

$\chi^2 = 12.67$

$p = 0.025$

d.f. = 4

FIGURE 35: HBV results by island of origin (antigen, antibody, negative)

I now knew that about 20 percent of people on the island had the antigen. Importantly, I discovered that there were no antigen-positive children younger than nine months of age. This information enabled my next step—figuring out how the virus was transmitted. I only needed to observe babies between zero and six months of age and follow them for a few months. I didn't need to study every child, just a subset.

Home for the Holidays

My return to America for the Christmas holiday was its own adventure. I caught the Islander from Santo to Vila only to discover that my flight to Fiji was canceled due to a cracked windshield. Air Pacific placed passengers in a resort hotel, Le Lagon, for two nights while the plane was being repaired. The airline sent a telegram to notify my mother, which she didn't receive. As planned, she drove the family to the airport, and everyone was at the gate to meet the Nadi–San Francisco Qantas flight. But . . . there was no Betsy on board. This was December 22.

Panicked, she returned home and dialed the operator. "I need to call number 510 in New Hebrides," Mom said. She knew Ken Hutton's number by heart.

"Just a moment, please, while I connect you," the operator promised.

After what seemed like an eternity, she heard a man's Australian accent on the line. "Ken Hutton."

"This is Marilyn, Betsy's mother. She was supposed to arrive today from Fiji. Have you heard anything? Is there something I should be aware of?"

"Marilyn, there's no news of plane crashes or accidents. I saw her a couple of days ago and she was fine. All excited about going home for Christmas. Leave me your phone number. If I hear anything, I'll ring you. Trust me, she'll be all right. She's a good Septic Tank."

Mom left her number with Ken but wasn't satisfied. She called my best friend, Dawn.

"Dawn, Betsy's missing. Her flight arrived without her. Her friend in Santo says there haven't been any plane crashes, so she must be alive. Any thoughts?"

"Well, she does have a weakness for good-looking men. Maybe she met someone who offered her a once-in-a-lifetime Christmas in Bali."

"She'd send a telegram. She's not that irresponsible or that inconsiderate not to tell us that her plans had changed," Mom said.

"Fair enough. If you want, I can look into it and see what I can find out," Dawn offered. "Give me her travel details."

Dawn was on the Case of the Missing Daughter. When she called Air Pacific, she discovered the truth. Everyone at home was relieved.

Back in Port Vila, we stranded travelers thoroughly enjoyed our all-expenses-paid vacation at Le Lagon. Most were Australians. I was the only American. We dined together in the evenings, traded stories about our lives over drinks, and looked forward to being with our families for Christmas. I left Port Vila for Nadi, Fiji, on Christmas Eve.

On Christmas Day, we took off for San Francisco. Approximately thirty minutes before landing, the captain informed us that the West Coast was fogged in and we would be landing in Las Vegas, Nevada.

Seriously? Christmas in Las Vegas?

A 747's worth of Australian passengers cleared customs in a warehouse at McCarran Field,[2] and Qantas put us up in a hotel on the Strip. That night, we met for dinner and wanted to order the most expensive dinner we could . . . Australian lobster! But that was too close for comfort, so we had hamburgers instead. The Aussies, most of whom had never been to America, came back to the hotel after a night of gambling and we debriefed at the bar.

"America is amazing. Drinks are free. Food is free," they enthused.

"Don't expect San Francisco to be like this," I warned. "It's very different. Vegas is unique in America." I don't think they believed me, and very few of them got any sleep, making certain to soak up all the hedonism they could.

The next morning, December 26, I arrived in San Francisco. My entire family came to the airport to meet me at the gate . . . for the third time. It was the only time in my life when I experienced two Christmas days in the same year, and I spent neither of them with family. The International Date Line made it possible to celebrate birth and rebirth within a span of mere hours.

I stayed in the continental US for about seven weeks and soon realized my fantasies would not be fulfilled. Even though I thought it was over with Marty, he flew out to see me in California. That sealed it. He had been a jerk, and the more of a jerk he was, the less handsome he became.

---

2. Now Harry Reid International Airport.

On my way to Philadelphia, I took a detour via Toronto to visit Michael, and I stayed in his apartment for a day or two. He said our exchange had been wonderful—it was intense and heartfelt, and we needed to take it to the next level. Nope. It turned out that things were different in person than they were in prose. I was content with the status quo, so things cooled off.

In Philadelphia, my meetings with Goodenough went okay, but they weren't as open and friendly as they had been before I left.

At ICR, nothing I did was good enough for Blumberg. He made it quite clear I hadn't wowed him with either my results or the number of samples I brought back. I knew nothing I did would ever match what he wanted, because he had to hold the reins. And I knew that for the work I was doing to be meaningful in his world, it needed to be done by a physician and not by an ignorant, untrained anthropology graduate student. The coup de grâce was his confirmation that he was withdrawing my funding.

My "Home Sweet Home" fantasies were crushed—I'd imagined I was going back to my "real" life, the one in my head, but that wasn't the case at all. I wasn't that Betsy any longer, so America couldn't give me what I needed. Although America wasn't what I'd thought it would be, it had been critical to come back, take a break, and deal with the real world. I still had work to do. But I didn't hurry back to Malo. It was the rainy season, so it made sense to wait. It also made sense to leave America with a clear head. I thought, *Okay, I'll show you and prove you all wrong.*

What is the best place for serious thinking and introspection? A beach. I headed westward in mid-February and took a three-week stopover in Honolulu, where Bill was doing research at the Bishop Museum. He met me at the airport with a traditional lei of leaves, and he introduced me to sashimi. One day, he said, "Put on your bathing suit. We're going to Makapu'u."

"Okay," I said. Little did I know this would be the first time I thought I was going to die. When we arrived at the beach, Bill told me more.

"Makapu'u is the best bodysurfing beach in Hawaii. The waves can get over sixty feet high."

I gulped even though I loved bodysurfing. It sounded high, but a six-story building wasn't *that* tall. "Sixty feet?"

"Yes. That's the distance between the top of the wave and the trough. It's an easy day though—only thirty-foot waves."

Relieved, I left my towel on the beach, ran into the water, and dove under the crashing waves. I was a strong, competitive swimmer with no fear of water. I swam out beyond the breaking waves and lined myself up to catch a swell to ride in. As I swam to the crest of the swell and watched the ocean contract, I looked down from the height of a three-story building at the white sand floor below. No water, no cushion—just hard sand. I was seconds away from being crushed.

I turned around and swam as fast as I could away from the breaking wave and into the relatively calm ocean. I watched what other people were doing. There was one important difference between me and them: They all wore three-foot-long fins so they had control while bodysurfing. I had only my seven-inch feet. I knew if I tried to bodysurf in, I would be killed or injured. What to do? I treaded water for the next thirty minutes, thinking and watching. Finally, I saw a path that would bypass the waves and get me back to shore. It required swimming to one side of the beach (ironically, near the cliffs) and sneaking around the swells onto the shore.

Forty-three years later, I stopped at Makapu'u. Much to my disappointment, the waves were less than two feet high and it looked so safe that I didn't bother telling my friend my story of near-death survival.

During my visit with Bill, I met several anthropologists at the Bishop Museum, one of whom I would encounter on Malo two years later while he was doing botanical reconnaissance with an old girlfriend of Chet's. Even in isolated Melanesia, it *is* a small world!

I left Hawaii, stopped briefly in Nouméa to visit my Santo friend Meythe, and then returned to New Hebrides. I was back on Malo by the end of the rainy season in late March, on the last of Blumberg's money and saving every penny.

With no funding, a new year, and more data to collect, it was time to up my game. I didn't know how I would do it, only that I had to do it.

Lessons Learned

- Spend time organizing and setting up a system that will show you what you need—it is efficient.
- Delays and detours can be fun and interesting.
- It is possible to have two Christmas Days in the same year.
- An intellectual friendship isn't enough for a romantic relationship; there needs to be a physical attraction as well.

- Some people will never be satisfied and will always want more.
- People are snobbish about degrees and often overlook a person's knowledge or ability to learn.
- Look before you commit—it could save your life.
- Salvation comes from careful observation.

Chapter Fourteen:

Changes—A Wrong Leads to a Rite

In 1976, I changed. I returned to Malo knowing I could make a dissertation out of the serology data I had collected and analyzed. Additionally, I no longer had Michael to turn to as an outlet, which forced me to absorb the ni-Tamambo culture and my surroundings. I allowed myself to become part of their culture. To take the next step, I had to immerse myself in their culture.

I told myself, *Okay, enough of the crap—let's get down to business. I'm going to learn the language and do all the things I should have done.*

It started with the clothes. In 1975, I'd worn pink corduroy jeans because I wanted everyone to know that I could not be pigeonholed. I was different, and I was an American—I was proud of that. When I returned, however, both the symbols and mindset had to change, so I transitioned to Mother Hubbards, the loose garment most women wore. I went to Johnny Lum's store in Santo and bought a couple of yards of material. One of the Malo women showed me how to make the dress, and another let me use her hand-operated Singer sewing machine. We ripped the material into skirt, sleeve, and bodice pieces. Two hours later I had three Mother Hubbards.

I put my Western clothes in a box and wore them only when I left for the States. It was a total negation of what had gone before. My dresses were patterned—my favorite was green with big flowers—and looked a lot like Hawaiian shirts. There was a colorful band of edging bric-a-brac around the neckline and the short puffy sleeves. The top was fitted and the skirt was full so you could squat and discretely stuff the skirt between your legs; it was also long enough that you could carry two or three good-sized yams in it.

In addition to changing my style of dress, I learned to weave. I made baskets from both coconut fronds and pandanus leaves. I was instructed on how to separate the strands that could be woven and then learned how to braid them into a functional basket. Weaving a mat from coconut fronds was much simpler.

To complete my transition, I gathered my food the native way and transported it in a woven basket with the long strap slung around my hairline and the basket resting on my back. This system distributed up to thirty pounds of food by using bone rather than muscle. The result? I could easily carry a heavy load over long distances.

FIGURE 36: Carrying "groceries." Women returning from the garden carrying baskets from their foreheads.

FIGURE 37: Returning from the garden. Family carrying food and bamboo down from their garden on the plateau.

I turned off my kitchen refrigerator and added a stone hearth so I could cook like an islander. (However, I kept my laboratory refrigerator running.) I also ate like an islander. Instead of eating the nice cuts of beef, fresh bread, and Western vegetables I used to purchase in Santo, I went to the garden and dug up yams. I learned how to cook and bake them—and taro—in my stone hearth.

I gorged on the local foods I loved: papaya (a tree grew outside my door), more than twenty types of mango, and *ngai-ngai* nuts. My salad green was *havira*, the island cabbage. I could also safely open and grate coconuts. With these new skills, I made *laplap*, the staple dish of the islands. It soon became an essential part of my diet. The method and recipe are as follows:

- Start a fire on a bed of basalt-like stones.
- Skin three or four roots from the garden (taro, yam, or manioc).
- Cut three or four banana-like leaves from the trees near your house; strip the back of the leaf vein to use as string.
- Place the leaf-vein "strings" on the ground; lay the banana-like leaves on top of the leaf strings so that they crisscross each other.
- Grate the root and place it into the middle of the crisscrossed leaves; knead the paste gently and shape it into a circle that is at least one inch thick.
- Fold the leaf edges over the dough; wrap the root dough like a package and tie with the leaf-vein strings.
- Remove half of the stones from the fire with bamboo tongs; stir the coals into the remaining stones and make into a concave bed.
- Add the package containing the root dough and cover with the remaining stones.
- Cook until the stones are no longer hot, approximately forty to sixty minutes.
- While the *laplap* is cooking, grate the meat of a mature coconut.
- Remove the stones, open the *laplap*, and squeeze the grated coconut, dripping its cream onto the *laplap*.

This is the basic recipe (see Figures 16-20).[1] I don't expect anybody to actually make *laplap*, although I did prepare this tasty meal once in

1. It can be modified by adding meat, seafood, or vegetables to the leaf package. Coconut cream is added before or after cooking depending on local preferences and traditions. Laplap is both healthy and delicious.

Philadelphia and again in California using parchment paper instead of leaves. The outcome? Most people thought it was wretched and just took a courtesy nibble. Only one person threw up.

"Going Native"

Within a month of "going native," I could understand Tamambo. I was serious about learning the language and told everyone not to speak to me in English or Bislama. I tape-recorded islanders telling stories and listened to them as I fell asleep. One morning, after listening to a recording while sleeping, I woke up able to understand Tamambo. It was like a language light switch had flipped on. I must have absorbed the sounds, words, and structure without realizing it. When my comprehension switched on, I tried to talk; depending on how people responded, I could tell whether I had been understood.

When their answers didn't parse with my questions, I'd try again using different words and hand signs until I heard a credible response. People helped by telling me what I should have said. With time, I learned to use the right words and grammar. By the second month, I was fluent. My immersion was complete when I began to think and dream in Tamambo.

In 1976, I began orienting toward the women and hanging out with them. In 1975, I had perceived that the men held all the power, so I wanted to prove to myself that they could take me seriously. In my Western mind, women, even those trained in professions, were less powerful and therefore, with few exceptions, less useful. I mostly knew European men in Santo, and all of them were expats. My confidants were women, but the men had all the status and money. Often, I had been the only woman participating in expat lunches and conversations.

I now found value in the Malo women. Again, it was the women who enabled my next research step. Voligo, Vira's wife, and Vomule, Jole's wife, taught me the mundane yet vital skills of daily living.

When I started spending more time with women, I was surprised by how much they had to teach. This said more about me than it did about them. Their knowledge was nuanced, and in many respects, they were the culture's bedrock. Without them, many things wouldn't happen. Women went to the garden every day, ensured that meals were made, looked after the children, and fed the circle-tusk pigs. In contrast, men would often be found sitting in

FIGURE 38: Voligo, Vira's wife. Also Vomoli Havo's mother.

FIGURE 39: Vomule, Jole's wife

front of the cooperative store, drinking beer and playing Last Card, a version of Crazy Eights.

Men worked in the garden when it was time to recondition it through slash and burn. They would fish, build houses, and make and transport copra. They socialized through cards or chatted with each other (*storied on*). The women also *storied on*, but in open grassy spaces where the children played. Women worked constantly, even when it looked like they were relaxing.

When I returned after Christmas, I was more aware of the women's role and work, and I appreciated that they were the ones who held the fabric of the society together. Malo Islanders lived in what is often described as a breadbasket economy. Fruits and vegetables were abundant. The land and sea nurtured protein. They didn't have to work hard for food, but they did have to do something.

(For a view of "downtime" see Color Plate 6: Traditional Malo celebration attire, and Color Plate 7: Boy pumping well water. The celebration attire was worn during the visit of the British district agent to inaugurate the completion of the freshwater lens well near the Mission House.)

After I went native, each month brought a deeper understanding of the ni-Tamambo and their culture. Initially, I was in limbo about many things. It felt like I was slowly peeling an onion. Importantly, I immersed myself in their culture, to the point where I considered the possibility of staying permanently. That would mean marrying an islander—any older single man—instead of going home at the end of the year. I went to town infrequently and no longer did anything that would make people gossip.

Partnering with Power

Even after realizing how powerful the women were on Malo, I recognized that one man, Thompson, was pivotal. Born on Malo, he was a tall, smart man in his late thirties and was married with four children. I had gotten to know him before I went back to the States, but on my return, when I knew the blood-collecting phase of my work was almost complete, I was both more relaxed and more culturally comfortable. Thompson and I spent a lot of time together, and he was the next person to whom I owed thanks for making my work successful. His English wasn't great, so we spoke in Bislama until I was able to speak in Tamambo.

Our work was complementary. He would help me get blood samples to test for malaria and hepatitis. I would send the slides to Vila for malaria analysis; if a person was infected, Thompson would receive the results, which identified the species of malaria,[2] and he would treat the person. Both Thompson and I would get the data.

There were real benefits to this approach. He was one of the few on the island who had both professional status and a place of business. People respected the dresser, and he was a key resource for health and well-being. He kept vaccination records, dressed wounds, administered shots, supplied antibiotics, treated malaria, and sent people to the hospital in Santo. He was like a nurse practitioner or physician assistant in the US.

Thompson had done a training course in Port Vila, but the details were sketchy. What I did know was that he drank a lot, which is why we always stopped off for wine at the French plantation after visiting the remote eastern villages. There was no intellectual conversation during a trip with Thompson. He was big, powerful, and fun. While he was known to be a drinker and a bit of a good-time guy when he was off duty, I only saw him drink when we were on the way home in the back of the Toyota pickup truck. It was comfortable to share wine at the end of a long day. His job had status, so he didn't jeopardize it by drinking while on duty.

I knew his wife. She was a nurse, and I think they met when they were in training. However, I never saw that side of her. Thompson went to the dispensary at the same time every day, as did I, so we would be there together in the morning, have a break of two to three hours at midday, and then return in the afternoon. That was the workday structure in Santo and on Malo. It was very laid-back. He worked his way around the island on a flexible one- to three-month interval that approximated a "schedule." I got to know the island because of him, and I also got my blood samples.

Watershed—Actually, a Wineshed

One day, during what would be my last blood-collecting trip in 1975, Thompson and I drove to Tanmial, where Big Nambas from the adjacent island of Malekula had been relocated in the 1930s. The village was relatively close to the French plantation. On this last blood-collecting trip, I bought

2. All four species were identified on Malo: *Plasmodium (P) falciparum, P. malariae, P. vivax,* and one case of *P. ovale,* which is usually found in Africa. *P. vivax* was by far the most common species.

wine when we stopped at the French plantation. By this time, I had tested Thompson's serum and knew he had the highest titer (concentration) of hepatitis antigen of anyone on the island. Regardless, I thought the risk that I'd become infected with HBV was very low, so we continued sharing wine straight from the bottle. There were two or three other people sitting across from us in the back of the truck. They were hitching a ride to Avunatari. As we sat drinking with our backs against the cab, things took an unusual turn.

Concealed by a blanket, I started holding his hand, and one thing led to another. I thought, *Well, that's quite something.* I also realized just how many professional barriers I had broken and boundaries I had breached. I had intruded into another's personal space. It was taboo on so many levels. I was an anthropologist, and I was there to do important work—to observe and research. I certainly should not perturb the local culture to that extent.

It wasn't awkward afterward, but I did wonder what on earth had pushed me over that line. The next day, he invited me to spend a weekend with him in Santo. I can't deny that the thought of it was both titillating and scintillating, but, after mulling it over, I concluded that it was a terrible idea. I pulled back completely and decided that I had to focus, focus, focus from then on.

Looking back on that incident, it was a watershed. I had the confidence to be the aggressor—which was new. It was the beginning of me taking actual control of my life. I had turned the tables around, and I had power. This was my rite of passage.

Lessons Learned

- Accept the local cultural norms.
- It takes at least a year to acculturate, leaving your native culture and language behind.
- Remove defenses so you are able to honestly see yourself.
- Women are helpful and useful even if they are not perceived as powerful. They have much to teach and can be the key to your success; don't ignore them. They work continuously, men episodically.
- A sense of security or accomplishment allows exploration of new areas.

- Go native to understand natives.
- Life is a rite of passage.
- Men do not suffer the social consequences of promiscuity.
- You make poor decisions while drinking wine.
- It is possible to recover from crossing boundaries.
- Learning to take control is not a simple, straightforward process.

Chapter Fifteen:

Revisions, Revelations, and Respect

The work continued after I finished collecting blood. Next, I needed to find out how and why infants became hepatitis antigen positive as early as nine months of age. To do this, I identified infants who were six months old or younger and asked their families if I could observe them and take notes. Everyone agreed.

These nineteen infants were located within an hour's walking distance of Avunatari, with some families up on the plateau. I would arrive at their homes before ten o'clock in the morning and start my note-taking on the hour. I wrote down everything that happened to that infant in a sixty-minute time frame—who touched or held it, what went into the child's mouth, where the child's hands went, what the child touched, everything I could see.

Young babies were generally in constant contact with their mothers or an older sibling. If they were lying on a mat, someone was beside them. Older infants—those who could sit—would pick up objects and put them into their mouths. In general, the mother and preschool-age children would spend the day in or near the home when I was there. The mother would prepare lunch, and sometimes a grandparent or relative would come over as well.

I developed a shorthand on the spot and typed up my notes when I returned home. My notes were typically eight single-spaced pages long and took a couple of hours to type. I did this work, tedious and time-consuming, for six weeks. I reached a point where I knew there had to be a better way to capture this information. But how? I did what seemed completely counterintuitive at the time: I stopped my visits and read one of the books I'd brought.

I opened Burnet and White's classic *Natural History of Infectious Disease*. Chapters Seven and Eight changed my research. Chapter Seven, "Susceptibility and Resistance," covered the elements of exposure and consequence of infection: influence of age, different routes of transmission, and intensity of exposure. Chapter Eight, "How Infections Spread," put everything together for me. These chapters gave me a framework for organizing the behaviors I'd recorded in terms of how a behavior could transmit an infectious disease. For example, when the baby put an object into its mouth, the route of transmission was potentially fecal–oral. This could be relevant if the object had been handled by someone with unwashed hands.

Stepping Back in Time

I'd made progress with my research and saw my path forward. Then a letter from Bill arrived saying that his plans had changed and he wouldn't be able to stop in New Hebrides. Instead, he suggested I meet him in Honiara. *Sure, why not? I've never been to the Solomon Islands.* It would be a chance to spend time in another Melanesian culture with an experienced anthropologist. A couple of days later, I was on the plane and he met me at Henderson Field. After the requisite hug and "great to see you," Bill and I headed to the car.

"What do you know about Honiara?" he asked.

"Not much, other than the Solomon Islands were important in the Pacific Theater. I assume the Seabees built the landing strip," I said.

"Actually, no. Henderson Field was built by the Japanese and finished just before the marines landed on Guadalcanal."

"Honiara is on Guadalcanal? As in, the Battle of?"

"Correct. It was the first major land battle between the Americans and the Japanese. The fighting went on for more than six months. Our win here was a turning point in the war. For the first time, Japan had to retreat. It was bloody and thousands died and thousands more Americans were wounded. The Japanese lost twice as many soldiers. It was bloody and brutal." World War II was important to Bill, who had served in the navy.

That night, we went for a drink in one of the bars. At tables that seated four to eight were groups of American men. "Those men," Bill said, pointing to the different groups, "served here. Only another marine who was here understands what it was like. They come here to remember those who died as well."

"It sounds like it's a way to relive and process what had to be a traumatic experience."

"It is. If you haven't lived through war, you can't even imagine what it is like. As much as your friends and family want to help, they have no idea. Being here is like therapy for them."

"That makes sense. I've never seen ex-servicemen in Santo."

I took a moment and tried to project how I would feel, at my age, going to war. I couldn't. I heard the emotion and depth of feeling in their words. What they had lived through was beyond anything I'd experienced or could even imagine.

On the way back to Bill's flat, we drove through the jungle. He would occasionally pull over to the side of the road and point out rusted planes with faded Japanese markings on their wings. The planes were pinned between branches and were being devoured by vines. I shuddered when I imagined the twentysomething pilots whose last sight was this tropical rainforest.

The next day, as a special treat, Bill took me to California, the village where Vouza lived. "Vouza is the most famous of the Coastwatchers," Bill said.

"Coastwatchers?" I asked.

"Natives who provided intelligence to the US Marines on Japanese activities on the island. In 1942, soon after the Americans landed, Vouza was captured by the Japanese. He was tortured, then tied to a stake where he was stabbed multiple times with a bayonet and left for dead. But he wasn't dead. When the Japanese left, he gnawed through his ropes. He crawled back to the American camp to warn them about the Japanese plan to attack Tenaru."

"Was his information helpful?"

"His warning gave our troops the edge, and because of Vouza, we won the Battle of Tenaru."

"I can't wait to meet him."

"We're here," Bill said as he parked the car in a clearing.

Vouza, tall and thin with a graying bush of hair, emerged from his home, an Airstream-style trailer, in full uniformed regalia. We listened to the stories he told in Pijin,[1] which he illustrated with pictures and medals. I was mesmerized as he and Bill caught up with recent stories and exchanged bygone memories.

"Why is your home called California?" I asked.

"I named it California to remember my visit to San Diego. I had a wonderful trip there and I wanted to honor my American friends," Vouza said.

1. Pijin is the pidgin English spoken on the Solomon Islands; it is a dialect of Bislama and Tok Pisin, which is spoken in PNG.

I left in awe that battle-ready marines had been able to inspire such courage, bravery, and loyalty in a foreign culture in such a short time.

As we drove back to Honiara, Bill pointed out a stretch of ocean next to the coastline. "That's Iron Bottom Sound. It's called that because of all the sunken ships and downed planes that ended up on the seafloor there during the Battle of Guadalcanal." Farther down the coast, he added, "Since the war, it's impossible to swim in the ocean here. The sunken ships have formed an artificial reef, which makes it a paradise for sharks. The number of shark attacks is the highest in the world. There's no safe place to swim in the ocean. They've even attacked people swimming in rivers near the sea. People in canoes aren't safe. The only safe places to swim here are freshwater areas that are walled off from the ocean."

Great, I thought. *Honiara actually has the sharks I was afraid of but have never seen in New Hebrides.*

The following day, we visited an expat friend of Bill's whose house was near the shark-infested sea. I went off on my own to explore and walked on top of a low stone seawall. The ocean was turquoise, and the sand was fine and white. The water was no more than twelve to sixteen inches deep and, more to the point, I could easily see for a hundred yards. So, I did what any twenty-seven-year-old would do: I took off my flip-flops and jumped into the warm water. I walked and splashed for a bit. Then I remembered Bill's warning. So, I hopped back onto the wall and walked on it for a while. Since I didn't see an aquatic form of death, I jumped back in. The fourth time I did this in-out-in sequence, I had just jumped out when a very long black shark slammed into the wall where I had been, then pivoted and swam away. I looked down at the feet and legs I was grateful to have. This was the second time I'd cheated death or dismemberment in the Pacific Ocean. Sometimes it is useful to heed warnings.

The Biggest Prize in the World

Bill supported my research and thought that it was coming together better than anyone had expected. On that positive note, I returned to New Hebrides. When I arrived in Santo, I learned there was a telegram for me at the post office.

"Shoot," I said to Heather. "I guess that means my grandmother died. She was ninety-eight, but I'd hoped she'd make it to one hundred."

I walked through town and up the hill to the post office to retrieve the telegram. I opened it, expecting the obvious. Instead, it read "Baruch Won Nobel Prize for Your Work. Congratulations. Love, Mom."

I shook my head in disbelief and, in a daze, meandered back down the hill and into town. Dick Baker, the BDA, pulled his car over next to me and stopped. "Are you okay?" he asked through the open passenger window.

Startled, I looked up and said, "Yes. I'm okay, but my adviser just won the Nobel Prize." Again, I shook my head in disbelief.

"Get in. I'll give you a lift into town," he said, and opened the passenger door.

"Thank you. I still can't believe it . . . the Nobel Prize."

"So your work *is* important," he responded.

When I returned to Malo that evening, I told Vira the news.

"What's that?"

"It's the biggest prize in the world," I said in Tamambo. I mounted the telegram on the bamboo wall of my outhouse next to a gorgeous picture of Mali, where studies had been done showing an association between a slow virus, HBV, and liver cancer.

Back to Business

Upon returning to Malo, I picked up where I had left off. I reopened *Natural History of Infectious Disease* and used the authors' transmission classifications to code my typed observation notes. I underlined behaviors that could transmit an infectious disease to the infant and noted the potential route of transmission. The result was a list of behaviors that could be grouped into four categories: droplet (airborne), arthropod (insects like mosquitoes or lice), fecal–oral (hand to mouth), and parenteral (exposure to blood from open wounds, or access to the circulatory system like through open gums in a teething infant).

I tested my new system using a different set of observations and then made refinements. I distilled the behaviors into a checklist, grouped by route of transmission, which I typed and taped to a long clipboard (see Appendix II). I practiced making observations and found I could easily record one event every twenty seconds. Based on no statistical rationale, I decided to do four hours of observation during the day, for a total of 720 recorded events during each visit. I met every one of my nineteen families multiple times in their homes, and every day-long observation began at 9:00 a.m., with additional hour-long observation sessions at 10:30 a.m., 1:00 p.m., and 2:30 p.m. The events I recorded were those that occurred at twenty, forty, and sixty seconds during every minute of each hour-long session. I took a half-hour break between the two morning observations and another half-hour break between the two afternoon observations. Lunch with the family gave me an additional ninety-minute break. At the end of the day, I took a blood sample from the infant and family members. This was my routine during the last three to four months of my fieldwork. Each infant had at least two observation days, and often more.

As my fieldwork progressed and the end of my stay approached, I was certain I had enough material and information for a dissertation. I didn't know whether the pieces would mesh to create a picture of the natural history of HBV. I thought it was possible, but before I left the island, I needed to address any loose ends. Bob had already gone, so I was on my own, and I was focused on what I needed to do. I could speak the language and was well integrated for an outsider. My Mother Hubbards were second nature by then. All my changes and adaptations had resulted in the islanders perceiving me differently than they originally had—and fortunately, much more favorably.

Growing Pains

As my fieldwork progressed, the links to my old life were changing. I didn't feel like the old Betsy.

For instance, I was no longer the Betsy who would drop everything and head to Oregon on short notice to attend her brother's wedding. In fact, I'd missed the wedding and he was angry, though he had no appreciation for how difficult it would have been for me to arrange. There was one flight a week out of New Hebrides. It would have taken three to four weeks for a round trip between Malo and Portland.

It must have been difficult for my family to process what I was experiencing. They rarely spoke about it, and, when they did, they had very clear notions of the godforsaken pit of uncivilized horror they thought I was living in. Blumberg never came to visit, despite his promises, and my stepfather, Sal, thought what I had been encouraged to do was criminal. According to him, I was being forced to live in a malaria-, elephantiasis-, and dengue-ridden cesspool with no preparation, putting my body at risk and potentially ruining my health forever.

"You're studying hepatitis!" he would proclaim. "How can they let you do that without protection?"

He felt it was unconscionable that any school would allow a young person to go into a disease-ridden area with no real education about infectious diseases or how to protect oneself. Ironically, he turned on the charm for Blumberg, writing a lovely congratulatory letter on his Nobel achievement.

I know my family was proud of me, although they rarely explicitly said so. During our scheduled telephone calls, they were relieved when the call went through, confirming I was still alive. My survival was their primary concern.

On our phone calls, my mother would often ask, "How are your savages?" She relied on the terminology she'd learned reading Freud's *Totem and Taboo*.

"Mom," I'd reply. "They're not savages, they're people." It would be years before she accepted this and changed her vocabulary. Her letters still concentrated on disease rather than anything overtly emotional. It was her way of showing that she cared.

Mom would call during my monthly trip to Santo. We agreed she would call on the first Friday, and I would be available to take her call that Saturday afternoon at three o'clock.[2] The call within Santo was routed by an operator. It was convenient because I could tell my new, Australian boyfriend, Rick, who worked for the phone service, where I would be and he'd have the call rerouted. More often than not, I took her calls at the Hotel Santo, a restaurant I frequented with Heather. I would spend the weekend in town, do some minimal shopping while I was there, and then return to Malo—my monthly routine.

Ocean Life

It was nice to talk to my family, but I was evolving into a more self-confident and independent Betsy than the person they knew. I had seen things they would never see, like in 1976 when I saw the most amazing phenomenon. During the full moon in early October, there was a massive run of *palolo* worms. The sea became alive with long, spaghetti-like, rainbow-colored worms that were spawning. Some were bigger and thicker than others. The worms broke in half and their tails swam to the surface to release either sperm or eggs. I joined the islanders who waded out onto the reef until we

2. Vanuatu is on the other side of the international date line. San Francisco was eighteen hours behind Santo time.

were wet up to our waists. They carried lit coconut fronds like torches in one hand while scooping up the worms with the other and dropping them into a woven frond basket. At first, I found the worms' slimy texture disgusting, but once I got used to it, it was fun. This predictable and important event was part of their lunar calendar.

The *palolo* worms (also called *eli*) spawn for a couple of hours once a year. The islanders' tradition is that the men go to their gardens and dig up a special yam the following morning. That evening, the *eli* are cooked on top of the yam. This special *laplap* looks like it is covered with gray vermicelli. The meal is unique and celebratory.

Vira said people either love *eli* or hate it. I tried a piece, and I hated it. The *laplap* was salty and tasted awful. I ate one small bite. The whole experience, however, was very cool.

The ocean and everything associated with it was central to the ni-Tamambo way of life. For me, it was initially a source of terror. But by the time I witnessed and experienced the *palolo* worm run, I was more laid-back. That lasted until I was swimming one afternoon and Vira's daughter, Vomoli Havo, started screaming "*Baheo!*" (shark) over and over again. She was at the water's edge waving and yelling as loudly as she could. To put it in context, *Jaws* had been filling movie theaters for months, and my mother was constantly sending letters telling me how to behave *when* I encountered a shark.

Your bathing suit, she wrote, *must always be blue or green, never brightly colored. It is important that you blend in with the color of the water. If a shark is coming after you, don't splash, just swim calmly away.*

Oh shit, I thought. I was about two hundred feet from the shore, in deep water beyond the reef. I remembered my mother's letter and, with a racing heart, swam toward Vomoli using a slow, wake-free breaststroke. I completed each stroke with my head above water to monitor Vomoli, who was increasingly animated and shouting. Now really scared, I increased my swimming speed with a splash-free overhead crawl, certain each flutter kick would be my last. *The shark will bite my foot off. I will either bleed to death or be the main course for a feeding frenzy.* There was no point in stopping or turning around to see or confront my killer. I would never have been able to see the shark with my awful vision. More to the point, I would lose time. Getting to shore was critical. I swam until my arms touched the coral and then stood up. Vomoli was still there as I walked out and thanked her. Safe. With adrenaline still flooding my body, I fumbled for my glasses, put them on, and looked out to the ocean, where I had just escaped death for the third time.

Dolphins. All I saw were dolphins. I'd survived, but my first thought was, *I missed my chance to swim with the dolphins!* Had the dolphins chased off the shark? I'll never know.

The sea wasn't the only place with animals. As I became more comfortable with the people, I assimilated and saw more of what was around me. There were spiders everywhere, along with rats, bats, mice, cockroaches, lizards, and snakes. Emaciated dogs scampered nearby looking for food scraps. Feral cats ate the mice and rats. There were beetles, bugs, bamboo mites, and mosquitoes. Every night, I listened to the mites slowly, methodically munching away at my house. I slept under a mosquito net and also read under it during the day. It was the one bug-free place in my world. Regardless of the time of day, there was always the incessant sound of insects buzzing.

I felt safe and secure on Malo. It never occurred to me that anything could change that feeling. But one evening, after I'd been drinking in my kitchen with some of the men, I went into my house next door to sleep. Hours later, I heard someone creeping on the crushed coral outside my bedroom window. A man. With no preamble, and certainly without niceties, he propositioned me in Tamambo. It was very uncomfortable, and I was scared. I lay in bed, in my pajamas, under my tucked-in net, with the open window six inches away. The man stayed underneath the window whispering, talking, and making explicit sexual requests for what seemed like a lifetime. He was drunk and kept saying, "Come on! Let's go have sex! I want to have sex with you."

"No," I kept hissing. "Go away! I'm sleeping."

I was afraid that he would come in. There was nothing to stop him—no Bob, no locked doors. I was vulnerable, and he knew that. He was there for what seemed like an hour. I thought, *Not again. I've cleaned up my act.* It was eerily similar to the event in Santo eighteen months earlier, when I'd first arrived. This time, it was worse—the man was not a stranger; he was someone I knew.

The next day, I saw the man I suspected and told him someone had been under my window. "I was very scared, even when he finally gave up and left," I said pointedly.

"That is a terrible thing to have happened," he agreed with a completely straight face.

Everything went back to normal after that. I was never as relaxed around him as I had been before, but I was confident that I'd put an end to the harassment.

Besides, there was work to be done.

Lessons Learned

- When you're overwhelmed with work, find a way to work smarter and advance the project.
- World War II had a long-lasting emotional impact.
- Mutual respect and honor can make heroes out of local tribespeople, save the lives of foreigners, and win battles.
- Make conscious decisions rather than just going with the flow.
- It is useful to heed warnings.
- Real or imagined, sharks are everywhere.
- The deeper you integrate into the culture, the more nuance you can observe.
- Oblique confrontation can resolve problems with nobody losing face.

Chapter Sixteen:

Milestones and Millstones

As I became more interested in and involved with Malo's women, I also became more aware of what held their community together: interpersonal relationships and kinship. On a typical day, women would walk to their gardens, either alone or with other women and preschool children, if they had any. They would spend a couple of hours collecting yams, gardening, and cutting firewood, then convene for a midday snack and chat. In the afternoon, at home, they spent an hour or so preparing meals, with another hour devoted to feeding the human and nonhuman animals. There was no technology, so there were few distractions.

Whenever Malo Islanders met each other, there was always a greeting based on their kinship relationship. (This practice still goes on today I believe.) Everyone was related either by blood or by marriage, and many were related by both. In the latter case, the kinship ties could be closer to or further from the speaker depending on how and through whom the relationship was traced. The interchange went something like this:

"Hello, my mother's mother's brother's son's son." (In other words, "Hello, second cousin.") This greeting emphasized the direct blood, consanguineal, male-dominated relationship. Alternatively, if there was a desire to accentuate kinship distance, the same person could have been greeted as follows: "Hello, my mother's mother's brother's son's wife's son." This greeting traced the relationship through a marital, or affinal, line.

The person being greeted responded with the appropriate reciprocal, either "Hello, my father's father's sister's daughter's son" or "Hello, my mother's husband's father's sister's daughter's son." Got it? It was stunningly difficult for me to keep track of, but they did. The interaction always ended with a sign-off indicating when they expected to see each other again—soon, tomorrow, or sometime in the distant future.

Their complex interpersonal dance was highly choreographed. Islanders were always talking to people and engaging with them. I would take alone time to write or read—but they didn't. Their days were filled with people, and they had highly developed person-to-person skills.

I got used to the high-contact personal element, but reaching that level was exhausting. In 1975, when I didn't understand the language, I suffered from sensory overload. But once I was fluent in the language, daily interactions became easier. It was like being in a nest—comfortable and full of support, if a bit crowded.

The focus on personal interactions was different from the American life I had led, where alone time was expected. I hadn't spent time thinking about how to interact with others when I was in the States. On reflection, my social interactions were primarily give-and-take rather than comfortable and supportive. By comparison, Malo was a womb of social support that was integral to islanders' daily existence and as natural as breathing. They spent their time and energy maintaining a cohesive social framework. I found it remarkable because it was unfamiliar. And yet the feeling of being secure and relaxed reminded me of the soothing and restorative effects of a warm bath.

It Takes a Village

I didn't see children squabbling on Malo. They played together nicely. I know this sounds idyllic, but I can't recall hearing a mother yell "No, no, no! Don't do this! Don't do that!" She might say, in an even tone, "Be careful." There was no helicopter parenting, and groups of children weren't homogeneous in terms of age, although they often were in gender. Routinely, older children would care for their younger siblings, cousins, or others in the group.

Women typically had eight to ten children; frequently, two or three would be under the age of five, so childcare had to be managed in a way that worked for everyone. As a group, the children might be collecting firewood or walking to the garden, but I was never aware of children fighting over something.

The adults and adolescents were mindful. If something happened and a child was hurt, they dealt with it. I didn't see the clashes between kids and parents that I was familiar with in the States.

I think some of this harmony was a function of critical mass. When you have many children and they are close together in age, the dynamic is very different from having only one or two. In that case, a parent focuses on keeping those precious few safe and alive. When there are eight or ten, it seems that a herd phenomenon dilutes parental attention and individual egos within the sibling unit. I'm not saying a family could afford to lose a couple of kids; rather, parents couldn't obsess about them. If you are one of ten, rather than one of two, it appears that sibling rivalry for attention or objects is less intense.

Typically, a woman of childbearing age would have a nursing baby on her hip. Another child would be toddling, and a third would actively play with cousins who lived in the same patrilocal compound.

Childcare was practical and simple. If an infant needed to pee, the mother would recognize the signs, disengage the sling used to carry the child, and hold the child over the grass or dirt. There was a different cue for poo; the caregiver would pick up some soft, broad green leaves, let the child poop, and then wipe their bottom, collect the mess, cover it with more leaves, and dispose of it. It was simple, uncomplicated. No looking for a bathroom with a changing station, no unfolding a disposable diaper, no special wipe, no diaper bag—none of that.

It was fascinating to watch the way children were raised. Women had babies until they reached menopause, and they also had their daily chores, which involved going to the garden and performing chores with both hands. They created a solution for carrying babies that was both simple and functional:

- Purchase two meters of cloth.[1]
- Secure one end of the cloth under the right armpit.
- In a counterclockwise direction, wrap the free end of the cloth securely around the baby's butt and back, which rests on the woman's left hip.
- Tuck the remaining cloth under the part held in place by the armpit.
- Loop the free end under the cloth on the shoulder (held in place

1. Traditionally, they would weave an *imbamba* (sling) from pandanus strips.

FIGURE 40: Typical family of six children

FIGURE 41: Natures' "wipe." Leaves are used to clean up poop.

under the armpit). The baby's weight keeps the free end secure so it doesn't slip and yet can be pulled free easily.

- Ensure the baby is secure on the hip and adjust the length or tension as needed.
- Let go—the baby's weight creates tension and keeps the fabric sling secure.

We make infant-related interactions so hard in the States. The Malo babies could look around and interact with their caregivers. This doesn't happen when the infant's face is up against the caregiver's chest or back in a pack. The same is true when the infant hangs suspended in space on the parent's front, where there is visual stimulation but no face-to-face interaction. On the Malo mother's hip, infants experienced 180 degrees of their world, both horizontally and vertically. This simple solution was safe, and the caregiver's hip bone bore the child's weight for its first two years.

Possibly as a consequence of them being held so much by their mothers/caregivers and/or being placed on the rough coral soil, I never saw an infant crawl. Their ambulatory phase, prior to walking, was on all fours, with the bottoms of their feet and palms flat on the ground. It looked like a walking forward squat. Culturally, there was an emphasis on walking, which routinely began around ten months.

Another practical child-rearing practice was premastication. The mother or caregiver would chew food and then give it to the baby, like a bird. Infants cutting their first teeth ate what everyone had at the meal. As a result, infants were exposed to a variety of foods and tastes. Food allergies seemed nonexistent. It was simple, practical, natural, and probably healthy.

Child-rearing in the Western world is more complicated. We have infant formula and bottled or packaged baby food that's often filled with preservatives and sugar. Malo women spent no time with a blender, and they didn't attend courses on weaning; they just shared their meals and went on with daily living.

Whether it was childcare, feeding, or toileting, I learned it didn't have to be complicated. There was a simple way to navigate child-rearing. For instance, no one died from eating food that fell on the mat—a bit of dirt wasn't of concern to anyone. When I realized that raising a child did not have to be time-consuming, I appreciated that I could still be a good parent if I adopted some of the islanders' ways.

FIGURE 42A: Sling (*imbamba*): Step 1. Adjusting the proportions.

FIGURE 42B: Sling (*imbamba*): Step 2. Making the "knot." Tying the *imbamba* securely with one hand.

FIGURE 42C: Sling (*imbamba*): Step 3. Securing the *imbamba*.

FIGURE 42D: Sling (*imbamba*): Step 4. Hands are free. Shading infant with a frond from the "umbrella" tree.

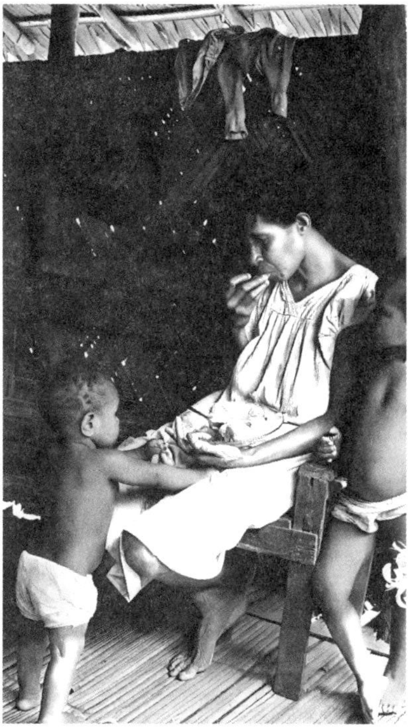

FIGURE 43A: Chewing (premasticating) food for toddler

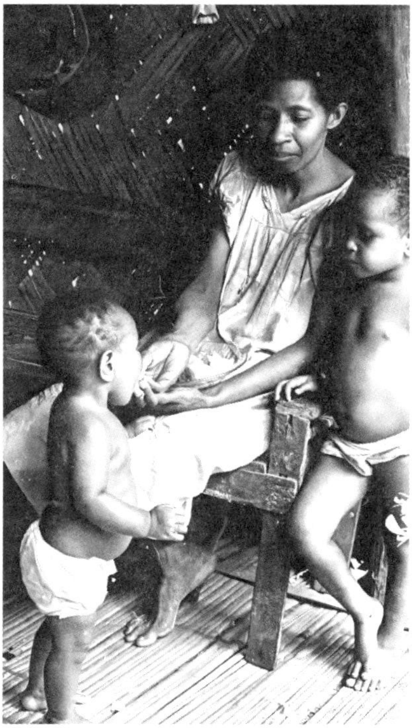

FIGURE 43B: Feeding premasticated food to toddler

Women often gave birth in the garden. It was not considered unusual, especially after the fifth or sixth child. They would work until they gave birth, tidy up the newborn, finish what they were doing, and walk home with the new family member. People were always nearby; help and support were a shout away. There was no multiday postpartum recovery.

For the first few deliveries, home was the preferred venue, since these births were riskier and more difficult. By the time child number four or five came along, however, the birth canal was stretched so deliveries were easier. There were also women who were recognized as helpers or midwives. Birth was seen as biologically natural and consistent with what other mammals experienced.

Dealing with Death

Life wasn't all idyllic on Malo. A woman could die during childbirth, and my genealogies identified many who had, although no one did while I was there. Childbirth was viewed as a natural process, and with that came occasional maternal or neonatal mortality.

When someone died—and hopefully, they'd really be dead—relatives and villagers would come to the deceased's home so they could see the body.[2] Mourners would bring blankets and lay them on top of the corpse, keeping the head visible. This wasn't just for symbolic comfort; it was also a token of investment. Each blanket was a gift to the dead. The number of blankets placed on a corpse varied depending on how many people came to mourn during the first day.

A line of mourners would form, and there would be wailing and weeping, one person at a time. When it was their turn, they would raise their hand and say, "I remember when we did [this]. I remember when we did [that]." Each person would recall how they felt: "You were wonderful" or "I'm going to miss you so much." The mourning was stylized, and it gave everyone present an opportunity to say something and have a moment to express their grief. The immediate nuclear family would sit around the body. They were encircled by other relatives, often based on how close or distant their relationship was to the deceased. The immediate family received support and comfort, and the wailing and outpouring of emotion was cathartic.

After everyone present had expressed themselves, the body would be buried, usually within a few hours of death—most people stayed for the burial. Some of the men would dig the grave. The deceased would be placed in the pit with the blankets. After that, the grave was closed. For the next ten days, the extended family would stay at the homesite adjacent to the grave and spend time at the gravesite whenever they wanted. Usually, they were alone while they communed with their loved one. The immediate family used this time to regroup and reinforce their ties. This was an important part of the healing process—a reassertion of the glue that holds the family together.

---

2. I witnessed a funeral where the corpse began to sweat. Lina had suffered a stroke and had been taken to Santo for treatment. According to her family, the doctors said she would be dead by the time they took her back to Malo. So they brought her back and held her funeral. When she started to sweat, the wailing stopped and someone said, "Ma mambo" (she breathes). They took off the blankets. Unfortunately, the stroke was severe and she died, for real, about five days later.

Friends and distant relatives brought food during this period. The family never cooked. On the tenth day, when the formal mourning came to a close, there was a feast to celebrate the deceased. Villagers brought food to the grave and slaughtered a pig or bullock. On the eleventh day, the villagers and extended family dispersed and went back to their usual routines. It was as if the community was saying "We have supported you during this intense grieving period. Now it's time to get back to normal life."

In 1970s American culture, not crying meant you were strong. We didn't want to view dead bodies in their natural state. If the corpse wasn't cremated, it was either embalmed (and only vaguely recognizable) or presented in a closed casket. We seemed to prefer a more remote and clinical approach to the deceased. I often think the downside to our approach was protracted grieving.

I preferred the islanders' direct and confrontational approach—it made sense. You're going to die. You can't stay alive forever, not even with toxic medicines or machines. You walk until you can't. Then you sit around telling stories and having people come to you. Then you die. It seems reasonable and natural, and I hope that's how I spend my last days. Thanks to my time on the island, I recalibrated many of my American conceptions and forged a Malo–American amalgam.

I was so impressed by the ni-Tamambo ritual surrounding death that I implemented key elements of it when my stepfather, Sal, died.[3] I moved into my mother's house and stayed with her and my brothers for the next ten days. I was surprised by how comfortable and right it felt to leave on the eleventh day. It was as if we'd had enough time to process his death and do what needed to be done. Nine days didn't feel right, and eleven days would have been too long. There is something about a ten-day period that works, and the islanders had discovered it.[4]

When Sal died, I took a week off from work, which acknowledged my stepfather's importance in my life. As a family, we had time to grieve, which allowed us to move on. Of course, there were many teary moments, periods of uncontrollable grief, and spontaneous outbursts of crying. But, I sensed

---

3. My mother awoke one morning to find that Sal, who was next to her, had died during the night. Ever the physician, she wanted to know what had killed him. She knew that the sooner an autopsy was performed, the better the results would be. As a practical woman, and since all of her children were at the house, she had her sons and my husband carry Sal's body out of the house and place it into the back of her Buick station wagon. She then drove the body to UCSF and had them put Sal's corpse into the morgue. It was a first for UCSF!

4. The mourning period for the death of Queen Elizabeth II was also ten days.

that the bulk of grieving for Sal and bonding with my mother and brothers was done. It was time to move on, and it felt good to have a defined endpoint. I was grateful to have been exposed to Malo's grieving tradition, which gave structure to mourning and made Sal's death easier to manage.

On Malo, wailing was a sign of intense grief, and the amount of wailing diminished over the ten days. With Sal's death, I discovered that grieving and crying are physically and psychologically exhausting. I didn't have the courage to be overtly emotional, because I was the stepchild and "the strong one." But, when I was alone, I cried and wailed—and I think it helped. In retrospect, having permission to emote overtly for ten days is healthy.

Open Doors

Life and death, the Malo way—so much of it made sense. Childbirth, child-rearing, and funerals were life-cycle anchors. I wanted to fill in the middle spaces and make connections at other places along that continuum. After a while, I sought out people even when I wasn't working. They would sometimes come to my house to chat, or I would go to theirs. I could turn up any time during their waking hours. There was always an open door, and it didn't result in social death, as could sometimes happen in the States if you showed up unannounced.

One downside of Malo's social open-door society was that no one kept track of time. No one had a sense of whether it was one o'clock or eleven o'clock. The islanders measured daytime by the sun, which could be difficult to sense indoors, in the bush, or even in the garden. I had a Timex wristwatch and measured time by the clock. Sometimes, I found their approach to time frustrating, especially when there was work to be done.

One day, for example, I asked someone to come to my house at three o'clock. I hoped he would arrive on time but suspected it might be substantially later. The only thing I knew for certain was that he would not be early. So, when three o'clock came, I waited. When three thirty came, I started to pace and swear. By four o'clock, I decided to do something constructive. I opened an issue of *The New York Times Magazine*, which Bob's mother had sent him, and started to write away for every catalog offered in the magazine. I had written about twenty checks, for amounts ranging from $0.25 to $1.00, when my three o'clock "date" arrived. I put my checkbook away and we went to work.

Even though it cost more to mail the checks than they were worth, I did reap a benefit from that wasted time. For the next four to eight weeks, my mail was filled with catalogs offering jewelry, clothes, and furniture. The catalogs provided entertainment during other times when I had to wait. In particular, a blue lapis-faced watch with a delicate chain-link gold band caught my eye. I wrote a check and put it in an envelope addressed to the vendor. It was in my take-to-Santo mail pile. One night, in a moment of boredom, I wondered what the lapis watch would look like on my wrist, so I cut out the picture and, with the help of saliva as an adhesive, plastered it onto my wrist. I was shocked. It looked terrible. I then cut out another watch from the same catalog and tried it on. The second cutout wore well. I tore up the first check and wrote a new one. That watch, a Cartier Ellipse, was my 1976 Christmas present to myself for completing my fieldwork and was my constant companion for the next thirty-eight years—a reminder of boredom's benefits.

With time, I got used to the ni-Tamambo rhythms and drifted with the flow. The pace of life was much slower than I was used to—more relaxed and easygoing. I would occasionally hear of someone having a spousal argument on Malo, but afterward, they would say, "That's just how he is." There were no value judgments and no sense of people bitching behind each other's backs. Similarly, land disputes were usually resolved without going to court. The islanders didn't argue about trivial matters, since everyone had pretty much the same things. There was never a question of you having more bananas or breadfruit than me because either of us could go out and get more bananas or breadfruit if we wanted it. There were no remote controls to fight over, no jealousy that someone had a more fashionable pair of shoes—nothing like that. Malo's culture was not based on material wealth.

With time and language fluency, I got to know most of the nearby villagers reasonably well—well enough that I had a sense of who was a good informant on childcare or marriage patterns or genealogy or custom practices. I could sit comfortably in a group and chat, listen, or just be present with no expectation of having to perform, entertain, or do anything. The pace was smooth. No one got huffy, and I could participate or not. I could be me. Just me.

Should I Stay or Should I Go?

The more I integrated myself into island life and focused on my work, the more pressing the need to make a critical decision became: Should I marry an islander or leave? I had completed my observations and knew I could return to Penn with something useful from my experience and data. But did I want to go back? It was a serious question.

Did I want to take that next step and get to know the Malo Islanders at a deeper level? Did I want to fully immerse myself in their culture, or was it time to pack up and go back to the US? I looked around to see whom I could consider marrying. There were a couple of men who were attractive and who would also be more than adequate spouses. I know that sounds cold, but romantic love wasn't a factor. This was purely pragmatic.

Of the two viable candidates, the only one I still remember was a customs officer and a clear frontrunner. However, I never made any advances toward either man. Rather, it was a thought experiment. I figured that if I married in, I would not continue doing research. I would live the life of a Malo woman. No Westerner had done that, and it wasn't long before I realized I would never be a Malo woman and couldn't live the rest of my life as one of them. I was also surprised that I had acculturated so thoroughly that I needed to make a decision about whether or not to become an island wife. (Remember, this was 1976. It was still a very racially torn world. Interracial marriages were rare in America.)

If I had taken that step, I don't know what would have happened. Would I have been completely on my own? I tried to imagine my parents coming to the wedding and buying some pigs for the occasion. What would the wedding have looked like? Who would have attended? I never took it to that extent. It didn't get past being a possible next option.

When it became clear I would return to Penn, I didn't know what the future held. Still, I was certain the next phase of my life was back in America. In Philadelphia, Blumberg had stopped my stipend. I realized the outcome was better for me when I was in control, even if the situation wasn't perfect or easy. When I called the shots, it was easier to live with the consequences. The next step was obvious—get my own grant and support myself.[5]

---

5. I was not awarded a grant from the National Science Foundation and had to write another grant application. In the interim, from the time Blumberg stopped funding me in early 1976 until I raised my own grant money. I also supported myself with money I had inherited.

I decided to apply for a predoctoral fellowship from the National Institute of Mental Health (NIMH). Since professorships in anthropology were few, I knew that if I got the grant, I could likely support myself as an anthropologist. If I didn't get it, then I would finish my doctorate and go on to medical school. One way or another, I would be able to support myself. The conflict with Blumberg over blood collection, money, and equipment had forced me to make decisions that were more practical than I might have wanted.

I further realized that finishing a PhD is a big deal. Writing a dissertation is a lot of work, and it is not a trivial undertaking. I was afraid I might have to analyze my data without access to the ICR's resources and without Blumberg's support. In battles with Blumberg, I usually lost. Whenever I pushed back, he would come down on me like a ton of bricks. If I got the grant, I would have independence. Hopefully, I would also learn to avoid these nonproductive interactions that were similar to those I'd had with my stepfather, Sal, and Paul, my first husband. For me, the cliché "the strongest steel is cast by the hottest fire" was true.

It was time for me to take the step that would determine my career and my future. On my manual typewriter, I summarized my results and wrote the application for the NIMH predoctoral fellowship. It was the first time in my life when I did something I knew I couldn't have done any better. There would be no excuses to rationalize failure. This time, I gave it everything I had.

I sent the grant application off to Blumberg and Goodenough, who were both favorably impressed. After this, my relationship with Blumberg improved and he became much more supportive of my research.

November 17, 1976, Tuesday, 9:30 p.m.—From Mom

Dear Betsy,

I left word yesterday for Dr. Goodenough to give me a collect call when it was convenient—he just called. He was very friendly, cheerful, and altogether nice to talk to.

What I got out of it definitely was that he and Blumberg approved—everything is available to you for work at the Institute. Goodenough was the [grant] sponsor and he and Blumberg are co-supervisors. They were pleased with the application—much better than what you had done previously and very reasonable—they xeroxed it, typed as it was, and rushed it off in time. He has no idea how NIMH will take to it—and he has no idea if it will go through—but he hopes it will. He feels if it doesn't go through you shouldn't be crushed.

I was pleased with and relieved by my mother's letter. Now all I had to do was wait nine months for an answer from NIMH.

Lessons Learned

- Spend more energy on people than on technology; also spend more time with people than in isolation.
- Critical mass makes raising children easier and is natural.
- Westerners make child-rearing harder and more complicated than it needs to be.
- When you know people, you learn how best to engage with them.
- Having ten days of grieving is important for processing the death of a loved one.
- Grieving time spent with your family reinforces solidarity.
- It is healthy to access your emotions and express them.
- Boredom, if used productively, reaps material benefits.
- It is very good to have an endpoint.
- Have a career plan B.
- You can be proud if you know you couldn't have done better, regardless of the outcome.
- People's opinions of you can change when you do good work and exceed expectations.
- You can usually trust your parents to act in your best interest with others.

Chapter Seventeen:

You *Can* Go Home Again

November 1976 was an important time for more than my grant application. Santo was preparing for a new election because the Nagriamel party had disputed the Northern District results of the November 1975 election. On Malo, the Presbyterian-backed National Party was pro-British. Up on the plateau, the Malo Islanders, who were more traditional or *kastom*, were members of the Nagriamel political party, which was supported by the French and led by Jimmy Stevens.

In the US, elections were also happening. I had been following the campaigns as best I could: listening to Radio Australia, reading the international edition of *Time*, and finding the occasional US-focused radio station. That first Wednesday in November, I sat in my kitchen, pinned to the radio as the results came in. I was thrilled when Jimmy Carter's election was announced, as it heralded the end of the Nixon era. I went over to Santo that weekend and brought a bottle of champagne to Heather and Ted's house.

"We won the election. The celebration is on me!" I said as I produced a bottle of Laurent-Perrier Grand Siècle, a favorite of theirs.

"But Betsy," Heather said, looking perplexed, "the Santo results haven't been tallied yet."

"The American election, Heather," I said. "Not the New Hebridean one."

"Ahh, right. Let's celebrate."

Heather graciously opened the bottle and poured the champagne. It was another weekend in Santo when I indulged in Western food, including my favorite: ice cream. By this time, I'd figured out that Rick, who was cute and funny, suffered from alcoholism. I thought it seemed excessive for him to drink a case of Foster's beer every day. But the clincher was the hideous, sickening odor of acetone. So, back to Malo for another four weeks and less time with Rick.

After I decided not to marry a local, I asked myself when would I go home—when would the time frame be right? I decided it would be December.

November 25, 1976

Dear Mom,

I thought that I'd write to say Happy Thanksgiving. I have the afternoon to do little or nothing. I just have to catch up on my lab work during the night and then I can go over to Santo on Saturday with a clear conscience. Actually, this week has been both trying and relaxing. It seems to be either of the extremes and rarely something in the middle. Right now, I am trying to finish off my observations, and with so few ladies left to do, when they forget and go up to the garden, it's quite disruptive.

Another of my favorite old men died yesterday. He died in the French Hospital in Santo, apparently from a huge septic sore on his leg. They took his body up [to the plateau] yesterday, and today most of the people went up to wail over him. This one I passed up as they had already buried him. He really was just the neatest man. When I would meet him on the road we would chat and laugh. During the time that I have been here, there have been more than the usual number of deaths and marriages, and perhaps even births. It seems that almost every family has a baby.

I have just about packed up everything, except those items I'm going to continue to use until the bitter end. It is starting to look a bit deserted. People are dropping in and talking about how much they are going to miss me and what they are going to do for me. I have just drawn up my list of who gets what.

About Christmas, I am very serious about not wanting presents, except for clothes! Don't rationalize any of this, just accept it and restrain yourself.

Remember, we all have to grow up sometime and I really would prefer to have Christmas this way. I know you're thinking that I may mean this now, but when I see the boys (God, adults) opening theirs, I'm going to feel left out . . . banish the thought. I know what I want! I hope this has convinced you.

As I look back on this past year and a half that I have been on Malo, I think that, especially during these last nine months, I have finally matured and actually grown up. I feel as though confidence has replaced arrogance, and I think I have much more control over myself and my feelings. I feel much better about myself, and I know that I can carry out a research project, and I can do problem-oriented research. I guess I feel as though I got my act together.

Apparently, we are now having our own Jaws here. According to the reports, a great white killed an eleven-year-old boy down on Rano Island just off of Malekula. From all descriptions, it sounds like the real thing. I think that I've gone swimming about once in the past seven months. Not from Jaws, because when I got back I was swimming every day, but now I just haven't had the time.

Well, enough for now. I'll be talking to you long before you get this. I'll give you the traditional collect call from Fiji to let you know that I made it that far. If anything happens this time that is unexpected, I'll let you know.

Love, Betsy

PS You know, the books are right. If you can go for a week without refined sugar, you really do lose your craving for anything sweet. So now, I just have starch lust (you know, taro, yam, biscuits, rice).

Just after sending off Mom's letter, I received a note from Blumberg; he agreed I'd collected enough material for a thesis presentation and expressed optimism about my chances of success. It was so positive that I felt we'd be able to make it work when I returned. It was an important confirmation.

Between making the decision to leave and departing the island, there were a couple of months during which I knew I needed to wrap things up and tie up the loose ends. That involved doing some things I had not yet done—going to Sunday church in Avunatari to see what it was like because most villagers went, making travel arrangements, letting people know when I would be leaving, and putting a bow on the relationships I had built.

Some of the families who were part of my systematic, quantitative observations lived as much as an hour's walk away, up on a plateau, so it was not trivial to go see them. I had to try—it was worth it. I needed to thank them for their help.

I was going to miss the people-oriented ways of the Malo Islanders. People only have a certain amount of energy, and in the America I knew, they focused their energy on work. (Nowadays, even more so as all our waking time is dedicated either to technology or activities.) On Malo, the ni-Tamambo focused their energy on personal relationships. I was going back to a culture with a very different orientation. For me, it was unexpectedly moving to leave these people.

And I had changed so much. Toward the end of my stay, I lost all sense of my skin color. Everyone around me on Malo was Black or Brown, and although I was white, I didn't feel different from them. I had a tiny round mirror I rarely used and had lost sense of what I looked like. You forget how much a mirror reinforces your sense of physical self. One thing I wasn't going to miss was having the children pinch my skin and fondle my hair. By the end, only an occasional child reminded me that I was different.

Saying goodbye was hard—I didn't know if I would be coming back. I assumed I would, but I didn't know when. There were a lot of goodbyes to a lot of people I had grown to love. They had been exquisitely generous, taking in a foreigner who was completely naïve and ignorant. They taught me their language and they let me become part of their community.

As I was saying goodbye to Vira, he said, "It's been fun having you, and please come back—but don't tell your friends about Malo, because you really were a lot of work!"

He was right. Think about it. Many communities would never do what they had done—or, if they did, it would take much, much longer. Allowing a total stranger to be integrated to the same extent was both incredible and generous.

I never doubted my decision, and as soon as I knew I was going to go back to the States, I focused on closing up. The date was set, and there was a lot of work to do. I had to seal the blood samples, making sure everything was packed up properly and ready to be shipped. There were lots of logistics to put in place.

The night before I left Malo, we had a party, a feast—the islanders slaughtered a bullock (not a pig), and we celebrated my time there. About forty people came. It was pleasant and relaxed, and there was no alcohol.

I would miss individuals. Some of the women I was sad to leave included my *kastom* friend, Vomoli Uri, who lived on the plateau and who had invited me to participate in a *kastom* dance one night. It was an honor to be included, and it demonstrated how much I was accepted. Vomule, Jole's wife, was my closest friend on Malo, and I would miss her humor and sensibilities. When I shared my upcoming departure with Voligo, Vira's wife, who had introduced me to the world of women, she said a strange thing. "You don't mind if we think of you as having died?" she asked. "We don't expect to ever see you again, so it is the same as death."

"Wait a minute! Being dead and being in America are two different things," I pointed out, but actually, that wasn't the case for them.

I don't know whether she mourned me, but we both teared up as we hugged goodbye—

I was very sad that I was leaving.

As the speedboat motored toward Santo, I looked back at the group who'd come to see me off. Uncontrollable tears streamed down my cheeks. I wept until they were out of sight. I had made my decision and commitments. There was no turning back. A chapter in my life had closed.

In Santo, I left my trunk filled with stuff at Ken Hutton's house. Heather graciously offered to find a boat heading to the US and have it shipped to San Francisco. About four or five months later, my mother went down to the wharf and collected my green footlocker from a Japanese fishing ship. That was it.

It was time to make my life and career back in America.

Lessons Learned

- In America, your finite energy is focused on work. On Malo, it's focused on people.
- Spend time making relationships with people work.
- Mirrors reinforce racial identity.
- Take the time and make the effort to tie up loose ends.

Chapter Eighteen:

Changes in Latitudes and Attitudes

I was armed with data when I arrived back in the States, and I knew I was starting a new chapter. I was also aware that I had closed a remarkable one. I had every expectation that I would return to Malo (which I did, four times), yet I still experienced a sense of loss.

I was also bringing a very different Betsy to the party this time, which made the transition back to America even more challenging. I remember sharing Christmas with my best friend, who is dynamic and funny—the very traits I admired. As my new Malo-ized self, I sat back and watched her perform.

Wow—she's quite loud. She doesn't listen, I thought. *It's all about her being the center of attention.*

I realized I had been a lot like her when I arrived in Santo and Malo. My ni-Tamambo immersion had changed me. Now I found the cacophony disconcerting and wondered how I could have missed this aspect of my friend for so many years. She was loud and hyperenergetic, whereas I felt disconnected. I came across as very subdued and out of it.

"You're being very quiet," she said, noticing the change. "Where's your spark?"

It was definitely a criticism, but I was okay with being quiet. I was happy watching. She was the spark; she was "on" all the time, and I had been like that.

Besides, I hadn't lost my spark; it's just that I was now much more introspective. I didn't need to compete any longer. I just wanted to be present, as I had been on Malo. We could all have been much calmer and more relaxed about Christmas.

The islanders' peaceful, minimal-stress lifestyle, coupled with their organic form of community, was something I would strive to find in the States. The ni-Tamambo have lived on Malo for hundreds of years, if not thousands. They'd had to figure out how to get along, since they couldn't just pack up and leave if things didn't work out. Instead, they'd developed ways to avoid and resolve conflicts. Their intellectual energy was focused around "How do I make this relationship work?" rather than "If I can't get along with my neighbor, I'll make his life miserable" and "If I don't like my boss, I'll sabotage and undermine them."

In the United States, we tended to allow interpersonal relationships to take a back seat, especially to technology, but on the island, successful interpersonal relationships were vital.

Split Screen

It was reentry time, even if my mind wasn't completely harmonized with where my body was located. Cognitively, I was split—I wasn't back in New Hebrides, but I wasn't in America, either. I looked at American life and was amazed by what I observed. Food was everywhere. No matter where I looked, there were stores, pictures, and words that communicated "Eat." In contrast, I had just come from a place where there were no stores. Food abounded, but there was no pressure to snack and drink and overindulge. In Philadelphia, there were constant reminders to eat, eat, eat, eat, eat. Everything was about consumption.

Another bizarre thing mystified me. Where were the Black people? Intellectually, I knew they were around, but I couldn't see them. On Malo, skin color was homogeneous, so I focused on people's faces, physiques, and affects. Skin color wasn't a descriptor there. Philadelphia, however, was not a homogeneous city, and West Philadelphia, where I lived, was a diverse and mixed neighborhood. It was only disconcerting when I approached it logically—I would say to myself, *I know there must be Black people here, but I can't see them, so where are they?* It took about three months before I was able to "see" skin color as a descriptor again.

I had become truly color blind.

A popular Jimmy Buffett song, "Changes in Latitudes, Changes in Attitudes," captured some of what I felt—with yesterday gone and wondering what lay ahead.

Wish *Granted*

Despite Goodenough's assurances and the positive letter from Blumberg, I was uneasy. I didn't know what to expect when I returned to ICR. I no longer had funding, and while I waited for news about my grant, I found a temporary part-time job with another anthropology professor. I also enrolled in organic chemistry—which made learning Tamambo seem simple—in case I had to apply to medical school.

In March 1977, I received a letter from NIMH saying I'd been awarded the predoctoral fellowship. I was funded. I would have a career in anthropology. Thrilled, I dropped chemistry immediately and, when my stipend started, I quit my job. I planned to regularly go to ICR and analyze the data I had accumulated. I tried, but my Malo experiences were still too close, too intense. I wasn't in the right mindset to look at the data, nor would I be for the next two months. I simply wasn't ready to take the next step. In the end, I needed a full six months from the time I returned to the States to recalibrate myself academically and personally.

Upon my return, I moved in with Bill, who was now separated from his wife. We lived together for three months, but that didn't work. Going back to my own space was much better.

"You want something I can't give you," he told me. "We should just move on." I thought that made sense, so I did. It was amicable. I moved back into my apartment in West Philadelphia and there was no looking back, no second-guessing. I had returned wanting to marry, have children, and live a more traditional life; those were important pieces of what I wanted, and they were not on Bill's agenda, which reinforced that it was time to move on.

Lessons Learned

- Reentry into one's own society is difficult.
- What was once considered "normal" is actually aberrant.
- Finding ways to resolve rather than escalate conflicts results in win-win situations.
- Americans are too food-focused.
- It is possible to be color blind. Seeing different races is a state of mind that arises from habituation.
- Focus on similarities rather than differences.

Chapter Nineteen:

Infection Connection

After returning to the United States, I was moving on—yes. But where to? And how? It was easy to say, but the path was not so straightforward. How was I going to analyze my data? What was I going to do with it after completing the analysis? Blumberg was an early adopter and had purchased a Hewlett-Packard mainframe computer that was located in the basement at ICR. Computers were still new, and data was stored on large reels of magnetic tapes. In some respects, the computer resembled a gargantuan reel-to-reel tape recorder. Two years later, ICR upgraded to a larger mainframe, a DEC PDP-11/70, that was shared throughout the institute.

Cyndie, a recent mathematics graduate, was assigned to work with me. She was an assistant to Blumberg's biostatistician, and together we (actually, she) framed and structured a database so we could do a different analysis after I'd entered the raw data into the computer. Each islander had their own record that included demographic, genealogical, and descriptive information that linked to their hepatitis B results. If the islander was an infant, the record was also linked to their observational data. Success would be measured by our ability to easily perform simple and complex statistical analyses. It took months of input and refinement—I had returned with a massive amount of data that needed to be coded and collated. Once the data were in the computer, we checked and rechecked them for accuracy. Trial runs showed whether the computer totals matched my hand tallies. Data management was not trivial, *and* I had Blumberg's full support.

From my perspective, one piece of information was missing: I didn't know which infants had become chronic carriers. I found that some infants had converted to being antigen positive while I was there, including the three-day-old infant whose toe I'd pricked. He was born with the maternal antibody and seroconverted to being antigen positive—a clear counterexample of the "maternal" effect. This was evidence that there were other means of transmitting HBV. But was he still antigen positive? Was he a chronic carrier of the virus?

There was only one way to find out.

Answers and Questions

I returned to Malo for a few weeks in 1978. My house and lab at Jole's place were hopelessly rundown, so Vira built a hut for me on his property at the beach. I collected samples from the nineteen toddlers and their families, processed these finger stick samples, and sealed them to return to the US. I caught up with my Santo friends (Ken, Ted, and Heather), and I was amazed at how much older and more accomplished Ted and Heather's five children were, all speaking French.

My new samples were analyzed at ICR, and they turned out to be a critical piece in understanding the natural history of infection with HBV. In fact, it came together quite beautifully.

Cyndie and I, with insights from Ed, the biostatistician, did the analyses together. It was fascinating—we could confirm a maternal effect using simple statistics, like a chi-squared test. If the mother was antigen positive, the child was significantly more likely to be positive. Simple enough.

But the analyses also revealed mysteries through contrary examples, such as the family in which both parents were chronic carriers but none of their four children was antigen positive. I wondered how they'd avoided being positive if virus transmission happened primarily through a placental leakage in the last stage of pregnancy and during birth. There was also an infant born with the maternal antibody who later became antigen positive. That shouldn't have happened—that child should have been negative, but he was a chronic carrier. There were so many questions. It made me challenge the simple hypothesis that transmission was due to the maternal effect.

When we combined the behavioral data with the data on which infants became chronic carriers, we could predict 100 percent of the chronic carriers. In addition, we could account for 100 percent of the negative cases. Our model was based on three important pieces of information: who was the source of the virus, which behaviors occurred during the source's interactions with the infant, and the frequency of those behaviors.

In other words, when an infant was sufficiently exposed and sequentially experienced three behaviors from an antigen-positive person during their first nine months, that infant became a chronic carrier of HBV.

These behaviors were:

- Exposure to droplet infection primarily through close face-to-face interactions when the infant began to smile at around three months of age.
- Blood or body fluid-to-blood (parenteral) exposure through saliva in premasticated food when the infant's teeth broke through the gums at around six months of age.
- Inter-person transfer of head lice, primarily while sleeping with the infant's head close to the caregiver's, generally around nine months of age.

In the case where the mother couldn't transmit the virus, the infant's virus-positive older sister transmitted it when providing these caregiving activities. She performed all the required behaviors with sufficient frequency such that her brother became a chronic carrier.

This explanation was powerful. Not only was it more comprehensive and nuanced than simply stating that there was a maternal effect, but also it showed the importance of collecting multiple blood samples over time to understand the natural history of a disease. What's more, these findings offered the opportunity for prevention by breaking one or more steps in the transmission chain,[1] which the transplacental transmission hypothesis did not.

I couldn't statistically differentiate which of these three behaviors was most important, but it was probably the premastication of food. This behavior varied the most. Everyone had head lice (and the head lice I brought back all tested positive for HBsAg) and everyone did "poochie poochie poochie" when an infant began to smile, but not everyone premasticated food. At

---

1. A summary of the cases is shown in Appendix IV. Stage I is when exposure is primarily insect-borne (mosquitoes, flies), Stage II is primarily exposure to the droplet route, Stage III is when parenteral exposure is more frequent through premasticated food, and Stage IV is exposure to head lice (insect-borne).

that time, there were public health promotions in favor of baby food and formula. The islanders were in a transition phase, and there were occasional clashes between traditional and modern ways, with premastication being in the traditional camp. Some of the younger mothers were moving toward baby food if they could afford it.

In any event, if all three of these behaviors occurred frequently enough with an infant who was antigen positive, then the infant became hepatitis B surface antigen positive and a chronic carrier of the virus. It was as if sequential exposure, through different routes of transmission, was needed to overwhelm the infant's maturing immune system. If any of these three behaviors failed to occur, then the child did not become a carrier. Being a chronic carrier was correlated with being at risk for liver cancer later in life.

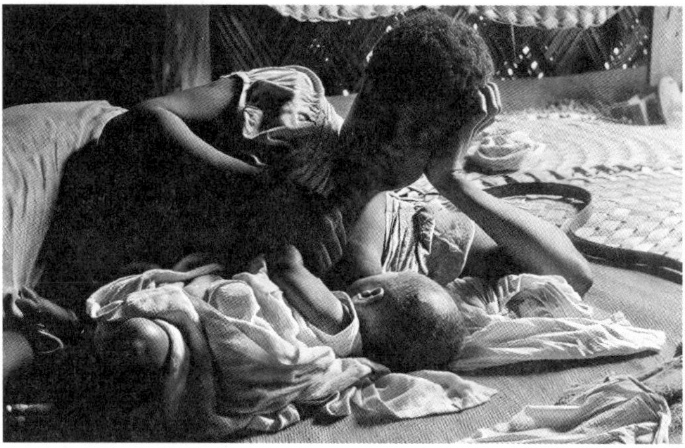

FIGURE 44A: Nursing mother—droplet transmission

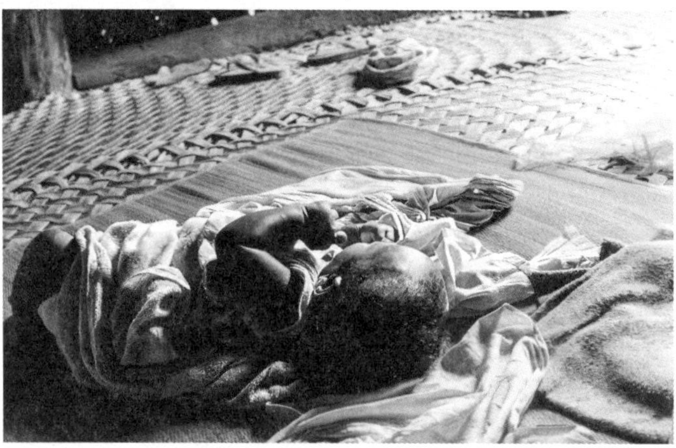

FIGURE 44B: Baby on mat—insect-borne transmission

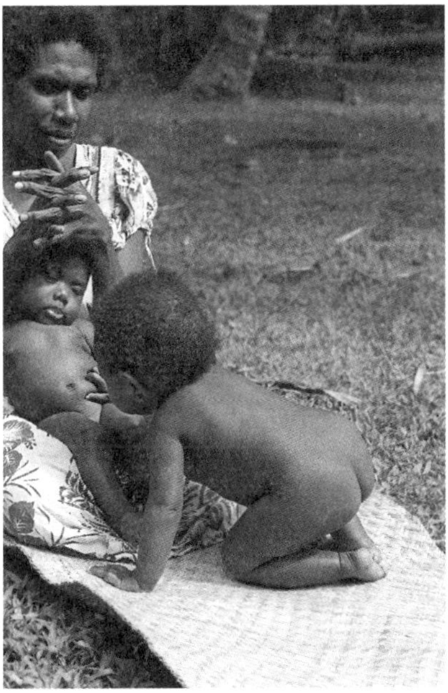

FIGURE 44C: Infants interacting—potential for droplet transmission

FIGURE 44D: Object in mouth—fecal-oral transmission

FIGURE 44E: Father-infant—droplet transmission

FIGURE 44F: Mother-infant—droplet transmission

FIGURE 44G: Looking for lice—insect transmission

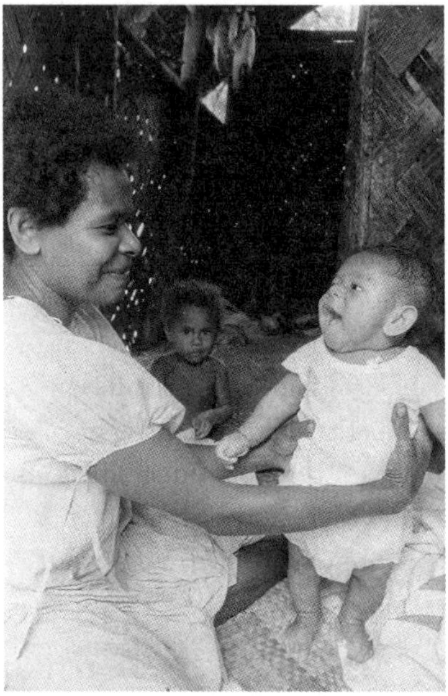

FIGURE 44H: Mother-infant—droplet transmission

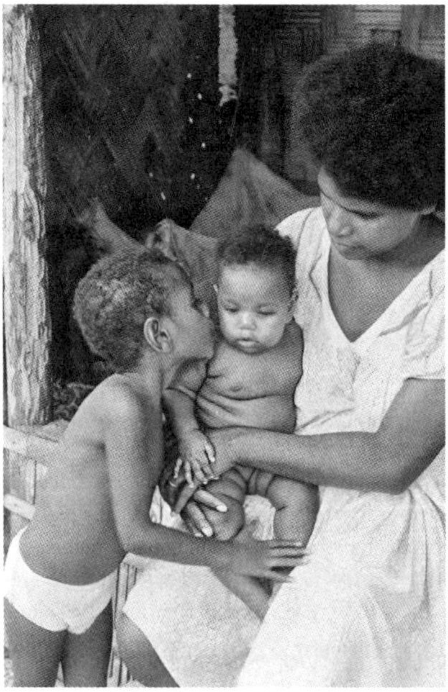

FIGURE 44I: Infant-sibling interaction—droplet transmission

The population of New Hebrides was so small—between sixty thousand and one hundred thousand people—that I didn't think there would be any liver cancer; the number of chronic carriers who develop primary hepatocellular carcinoma (PHC) is small, and it was thought that an additional risk factor, such as exposure to aflatoxin, was important.[2] Furthermore, PHC can take thirty, forty, or more years to develop and the average life expectancy at the time was fifty-seven years of age.[3] Hepatitis B is a slow virus that takes decades before its lethal effects are manifest. Statistically, one would expect to see only a few cases in the country. Based on the relationship between HBsAg and liver cancer seen in Taiwan, one would have predicted fifteen cases of PHC in Vanuatu. I was shocked and saddened to learn that Thompson, the dresser, was one of those cases. He died in his early forties.

Biologically, his cause of death made sense. He was a chronic carrier with the highest concentration of hepatitis B I had ever measured. His death confirmed that liver cancer occurred in New Hebrides. This observation suggested that chronic infection with HBV led to liver cancer regardless of whether the child seroconverted during their first year of life (New Hebrides) or after their second year (Taiwan). Although liver cancer was rare in New Hebrides, people were still at risk.

Lessons Learned

- Longitudinal data may give different answers to important questions than a single snapshot in time.
- Much of epidemiology is based on demographic statistics that identify association.
- The natural history of a disease is more nuanced when culturally relevant behavioral data are included.
- Behavioral and longitudinal data are difficult to collect; doing so requires in-depth cultural knowledge and an extensive time commitment, but it can lead to a deeper understanding of disease transmission and more relevant interventions.

---

2. Aflatoxin is produced by fungus found in warm and humid parts of the world and on agriculture crops, especially corn and peanuts, neither of which were present in New Hebrides. Aflatoxin is toxic to the liver and is also associated with causing liver cancer. Mark Feitelson, "Hepatitis B Virus Infection and Primary Hepatocellular Carcinoma," *Clinical Microbiology Reviews* 5, no. 3 (July 1992): 275–301, https://www.ncbi.nlm.nih.gov/pmc/articles/PMC358245/pdf/cmr00040-0069.pdf.
3. "Vanuatu—Life Expectancy at Birth, Total (Years)," IndexMundi, https://www.indexmundi.com/facts/vanuatu/indicator/SP.DYN.LE00.IN.

Chapter Twenty:

Politics, Profits, Schemers, and Scammers

The couple of weeks I spent on Malo in 1978 were pivotal not only to my work but to later events in the country.

In Malo's past, fertility control had not been relevant, because so few children survived to adulthood, but now that more than 90 percent of children survived, birth control was becoming important. The culture and land system were designed to support two to four children per family, not eight to fourteen. Birth control pills were increasingly available. All was good until a woman took me aside and showed me her arm. She had three rods under her skin.

"They told me this was good. I would not have babies for five years," she said. "I don't feel the same."

"What's it called?" I asked.

"Depot something," she said.[1]

As far as I knew, this product was not available in the US, and it looked as though this remote village was being used as a test population for a birth

1. The literature shows that in the early to mid-1970s, companies were experimenting with sustained delivery of contraceptives. Small-scale clinical studies were being conducted in the mid-1970s using Silastic rods containing levonorgestrel. See S. Roy et al., "Clinical End Endocrinologic Study of Continuous Levonorgestrel Administration from Subcutaneous Solid Polydimethylsiloxane Rods," *Contraception* 21, no. 6 (June 1980): 595–615. The levonorgestrel implant (Jadelle) received FDA approval in 1990 for medical use. In 1996 the product was approved as effective. In 2002 and 2006, the product received approvals for efficacy of three years and five years, respectively. These rods are a form of depot drug delivery and it is very possible that the word *depot* was used to describe the implant.

control implant. This type of birth control wasn't approved for use in the US until after 2000. Some health organization was clearly experimenting here by putting in a five-year implant to prevent pregnancy, and that was hugely problematic.

In fact, I was stunned that it had happened at all—it was tacky at best, and probably unethical. Women would show me their arms and I could see the rods. I thought, *Really—five years of that?* There was no obvious reporting or monitoring protocol—at least, the women did not describe any follow-up. The islanders were being used as a trial population. This was my first exposure to a pharmaceutical clinical trial—and it would not be my last.

Seeking Independence

When I was on Malo in 1978, I felt the same emotional connection with the people as I had before, but the country was more politicized. In 1975, they'd held their first elections as an initial step toward independence. What would independence look like? There were differing visions that aligned with the colonial Anglophile- or Francophile-backed political parties.

The National Party, originally British-backed, was supported by the World Council of Churches. It renamed itself Vanua'aku Pati (which translated to "my homeland party"). Among the Francophile parties, one of the largest was Nagriamel. It was funded by American businessmen and tacitly supported by the French government. Over the two years since I'd left Malo, Nagriamel had become increasingly aggressive and was campaigning for a separate, independent Espiritu Santo. The island had at least two other political parties, but Nagriamel was the most prominent.

Why was independence coming to a boil in the late 1970s? I've often thought it was an inevitable progression as the European colonial era was drawing to an end. After World War II, many African and Asian countries gained independence, some peacefully and others through bloody conflicts. Over the next thirty-five years, Britain divested the majority of its colonies, including its Pacific Island colonies.[2] Talk of independence from Britain and France was commonplace on Malo in 1978.

There were other forces, besides the British desire to leave the complicated condominium setup (a situation where two sovereign entities—in this case,

---

2. Fiji (1970), the Ellice Islands (now Kiribati, 1979), Nauru (1968), British Solomon Islands (now Solomon Islands, 1978), and Tonga (1970).

the British and the French—share control), that were driving New Hebrides toward independence.[3] Greed was one motivator. New Hebrides, like many other island colonies, had no direct taxation. As a condominium, there were no nationality restrictions on who could participate in this benefit. Beginning in 1970, individuals and corporations flocked to the country to take advantage of this unique situation.[4]

Between 1974 and 1978, prospectors came with the hope of finding a treasure of mineral resources. Unlike Bougainville's copper or New Caledonia's nickel, the best New Hebrides had to offer was aluminum, and not enough to make bauxite mining profitable. The only available resource was the land itself.

During my first visit, I'd heard about Eugene Peacock, an American real estate developer who came to New Hebrides in the early 1970s. He purchased a large tract of land from a Frenchman on the eastern coast of Espiritu Santo. Adjacent to his purchase was Champagne Beach, one of the most beautiful unspoiled beaches in the world. The light yellow-orange sand is fine-grained and there's a fringing reef a couple of hundred yards from the shore, which makes it safe for swimming. Peacock built a small hotel at Lonnoc Beach,[5] and he sold parcels to disillusioned American soldiers returning from Vietnam and residents of Hawaii who wanted to own a piece of beachfront paradise. Ultimately, the perceived threat of thousands of Americans living in New Hebrides was too much for the condominium government, so they passed a retroactive law that forbade selling property to American speculators. Peacock and unsuspecting buyers lost their investments.

Peacock was followed by another, more ambitious American, Michael Oliver. Oliver was scouring the world for a country he could make into a capitalist and libertarian utopia. He first came to Santo in the early 1970s but soon moved on to the Minerva Reef two hundred miles from Tonga. He claimed the reef, which was barely visible at low tide. I assumed it was like going to the lost and found. Oliver "found" the reef and proclaimed the Republic of Minerva to be a tax haven, presumably with himself as its supreme leader.

---

3. John Beasant's book *The Santo Rebellion: An Imperial Reckoning* (University of Hawaii Press, 1984) provides an excellent account of the history and behind-the-scenes activities, beginning with the formation of the Nagriamel party through Independence.

4. See "New Hebrides, in South Pacific, Is Becoming Popular Tax Haven," *New York Times*, May 10, 1971.

5. Then, it was called Lokelai Beach.

His next step in country formation was to dump tons of sand on the reef so it was just above sea level at high tide. The Tongans, unamused, asserted their claim to the reef as part of their traditional fishing area. In the "battle" of the Minerva Reef, the Tongans sent a ship carrying their king. They planted the Tongan flag on the raised reef, claiming and successfully establishing ownership. Unfortunately, a Tongan sailor fell overboard and drowned—a battle casualty.

After the Minerva debacle, Oliver shifted his focus to Abaco Island in the Bahamas, which had just become independent from Great Britain. Oliver's aptly named Phoenix Foundation aggressively supported the secession of Abaco. This, too, foundered and ended badly. By the mid-1970s, the Phoenix Foundation had risen for the fourth time and was back in discussions with Jimmy Stevens about an independent Espiritu Santo.

The smell of independence was in the air. Many thought that it could happen in 1980. The atmosphere was different in New Hebrides, and political discussions were frequent—only among the men, of course, when they played cards or sat along the edge of the street in Santo or in front of the Co-op on Malo. There was also overt discord among islanders as they discussed party affiliation—a new phenomenon.

Lessons Learned

- Pharmaceutical companies use Indigenous peoples to test new drugs.
- Foreign and capitalist interests can be disruptive to traditional values.
- Cultures change quickly.

Chapter Twenty-One:

Audacious Advances in Academia

In the spring of 1979, I'd had enough of the influences of politics and capitalism, but surely I was soon going to be immune—after all, I was looking for a job in academia.

There were few jobs to choose from; in anthropology, there were only four tenure-track assistant professorships available in the entire US. Tenure-track meant there was a career path to the ranks of associate and full professor if you obtained grants, published, taught, and jumped through all the hoops the university demanded.

There were also nontenured instructor positions. These were pay-for-hire teaching jobs that universities could terminate at the end of a semester. My goal was to secure a tenure-track position because they were guaranteed and offered a decent salary with benefits. If hired, I would have a future at that university or could use it as a stepping stone to another university.

During the application window, I was still analyzing my data, so I had not written my dissertation. As a result, I did not qualify for two of the four tenure-track positions, since they required a PhD. That left me with two possibilities: the University of Illinois, Urbana-Champaign; and the University of Texas, Austin. Both were good schools, but Illinois was slightly better because it was among the top ten universities in the country at that time, according to college rankings.

I applied and was invited to interview at both universities. I interviewed first at Texas. Some of the questions were worrisome. For instance: "You're a single woman—what if you get married? Will you still be driven? What would keep you from moving with your husband if his job wasn't in Austin?"

Yes, I was single. Why did it matter? How would I know whether I would move? The faculty interviewers made a lot of assumptions that made me uncomfortable. In addition, there were overt politics within the department as well as between the physical anthropology and cultural anthropology faculties. A professor whose children I'd previously babysat for was a full professor in cultural anthropology. I liked him and his wife a lot, which helped make the position attractive. However, it wasn't an environment where I wanted to work. So, I withdrew my application.

The interview at Illinois wasn't exactly straightforward either—but for different reasons. I got a little too comfortable at the party after my talk (maybe one or two too many glasses of wine) and started chatting with a prominent South American archaeologist.

Back in Philadelphia, my flamboyantly gay neighbor, Don,[1] had introduced this same short, white-haired archaeologist to me in the hallway, adding, "He's here for the weekend. I'm introducing him to gay sex."

"Great. Nice to meet you," I said, and went into my apartment. The next day, they came over for coffee and we had a brief conversation that included some discussion of Peruvian archaeology.

I had expected to see the archaeologist when I showed up for the interview at Illinois, but I had not anticipated his visible shock at seeing me. I could read his thoughts: *Oh my God. She knows I'm gay—no one else here does.*

Professor South America was in his fifties, married, and desperate to keep his Philadelphia assignation a secret. At the party, we started to chat, and then we both sat on the dining room table, where the drinks and glasses were. My Cinderella moment occurred when I swung my legs back and forth and one of my shoes flipped off. A South Asian professor picked it up to return it to me. At that moment, the table collapsed. Professor South Asia grabbed my hand and pulled me out of the crash, leaving poor Professor South America lying in the mess of glasses and wine bottles, probably asking himself how his life had reached such a low point and thinking, *Who is this new woman who just appeared, got tipsy, knows all my secrets, and broke the department chair's Danish modern teak table, into the bargain?*

The dean of the college repaired the table, and I was forever known as the Candidate Who Broke the Table. It didn't do me any harm, luckily, and

1. Don sent me an unsolicited copy of *Playgirl* while I was on Malo. Vira discovered it when the WHO sprayed my house while I was back in the States over Christmas. His only comment was, "*Man em he all same.*" I had forgotten about the magazine and was sufficiently embarrassed.

I got the offer letter in mid-April, contingent upon having graduated by December.[2]

Unwanted Advances

So it was settled. I had secured a tenure-track professorship at a Top Ten university. All I had to do was finish the data analysis, write my dissertation, have it reviewed, defend it, and get it approved by mid-October to make the December deadline. In addition, knowing I'd be "free" for the last quarter of the year, I secured a research postdoctoral fellowship at Stanford starting in October. So, after two and a half years of analysis, I wrote my dissertation during the month of August, committing to produce one chapter a week. I submitted a five-chapter dissertation, which I turned in just before Labor Day, and celebrated the long weekend on the beach at Cape May.

Up until August, things had gone well with Blumberg. I had access to ICR's resources, which were generously given while I supported myself with my NIMH fellowship. However, when I was on that tight timeline to finish my dissertation, it became more complicated. He insisted on reading each chapter with me sitting beside him in his office. He would put his hand on my thigh, and I would choose that moment to get up and "stretch." He even suggested that I come to his house for the best daiquiri in the world while his wife was out of town.

Really? Seriously? If this is what it took to get my dissertation approved, his approval wasn't worth it. I had the draft in hand and, with Goodenough's blessing, could easily change the composition of my committee. I didn't confront him, but I did a lot of standing and shuffling in his office while he read.

An affair or sex with Blumberg was not going to happen, and it never did happen. Not at the hepatitis meetings, not when he visited Stanford, not when I drove him to his hotel. Never. I wasn't so direct as to say "Not on your life," but I would move away from him, say I was busy, and tell him I couldn't possibly make it. I'd make all the excuses women have been using forever in the hope that the man in a position of power would get the message without it having any negative side effects.

---

2. My doctorate was awarded in December 1979. My dissertation was "Family Behavior and the Transmission of Hepatitis B Virus in Malo, New Hebrides: Ethology, Ethnography, and Epidemiology in the Study of the Natural History of Disease" (PhD diss., University of Pennsylvania, 1979).

Blumberg occasionally continued to try during my postdoc as well. I always rebuffed him, and I think that reinforced my resolve to have my career my way. From 1976 onward, I based my life on what *I was going to do* and on what *I could do*.

Before I knew whether I would be offered a university position, I'd explored doing postdoctoral work. I thought it would be fun and convenient to be in Manhattan. My stepfather knew microbiologist René Dubos, who was then at Rockefeller University, and arranged an interview for me. New York was attractive for many reasons, including its proximity to Boston. I'd met a cinematographer when a Boston-based WGBH crew filmed a *Nova* episode about Blumberg's odyssey of making the association between Australia antigen and HBV.

FIGURE 45: Baruch Blumberg filmed by WGBH for *Nova* (PBS)

Blumberg, however, wasn't keen on my going to New York and thought Stanford would be a better place for me. Besides, that meant he would have someone to play tennis with when he visited the adjacent Center for Advanced Study in the Behavioral Sciences. So he provided a great recommendation to his colleague, Luca Cavalli-Sforza, a well-known human geneticist and Stanford professor. This was why I spent the last part of 1979 at Stanford working with Cavalli-Sforza and Marcus Feldman, who were then studying the cultural transmission of traits. Cavalli-Sforza was at the height of his career in the medical school and Feldman was a theoretical biologist and mathematician in the biology department.

I was in California for three months before I moved to Champaign in January to teach biomedical anthropology. At that time, Urbana-Champaign was home to a Kraft factory and not much else. When I learned that the date for New Hebrides Independence would be July 30, 1980, I thought it could be fun to return, do some work, and participate in the festivities. I was dating Boyd, the *Nova* cinematographer, at the time, and he agreed. He even offered to film the Independence festivities if I could fund his airfare. I applied for money from the university and was awarded travel funds for the two of us. Boyd found a sponsor who would pay for his time, his film, a sound person, and other assorted expenses.

Returning to New Hebrides for Independence celebrations promised to be an exciting time of flitting from party to party. However, it became far more of an adventure than I could have imagined. It was my first time on the front line in a war zone.

Lessons Learned

- It is possible to write up four and a half years of work in five weeks.
- When turning down men in power, it must be done so they don't lose face.
- Stick to your values. Trading sex is rarely the easy way to achieve your goal.
- Don't reveal secrets; they can come back to haunt you.
- If you're the right person for the job, you can get tipsy and still get the offer.
- It doesn't matter where you want to go for your next step if your adviser doesn't support it. You'll end up where they want you to go.
- Universities are sexist, and they are full of powerful, manipulative people.

Chapter Twenty-Two:

War Zone Escapades

Other than New Hebrides gaining independence, it seemed as if there was little else happening in the world during May, June, and July of 1980. The US had decided to boycott the Moscow Olympics, and there was more of the usual and no longer interesting "turmoil" in South Africa. As a result, the major papers in the Francophile and Anglophile worlds sent reporters to Port Vila to cover the events leading up to Independence. The island and its people were picturesque. Photos of men wearing penis wrappers and holding bows, spears, and single-action carbines (rifles) made great copy and sold periodicals.

Jim Biddulph published an article in *The Listener* describing the scene:

> Take a few dozen islands in the Coral Sea, mix in active volcanoes and typhoons; a local population much reduced by disease and forced labor but still of significant size; rival and mutually critical British and French administrations; a sagging economy, and independence only a few weeks away. Garnish with three legal systems; exotic cults which may or may not occasionally involve cannibalism; sundry secessionist movements; right-wing American speculators; British administrators poised for flight; French settlers scheming to hang on in the name of French culture; Australian adventurers rollicking round the fringes. Welcome to the New Hebrides.[1]

1. Jim Biddulph, "Squabbles in the South Pacific: The British and French Will Leave No Legacy of Cohesive Government to the New Hebrides," *Listener*, June 26, 1980, 816.

The cosmopolitan reporters also found Bislama (pidgin English), which is written phonetically, to be a source of amusement: "New Hebrides officially renounced cannibalism a long time ago. But one can never be completely certain. Even the first three words of the new national anthem sound sufficiently evocative of a cannibal festivity; *Yumi, Yumi, Yumi*."[2]

I was stunned to read in the newspaper that there had been a coup d'état on the evening of May 27 with the secession of Espiritu Santo from the New Hebrides. The eye-catching headline was "Bow-and-Arrow Rebellion."[3] Jimmy Stevens announced the new country's name, Vemarana. The article also cited the nascent chief minister, Father Walter Lini, and British high commissioner, Andrew Stuart, saying New Hebrides—with Espiritu Santo—would become Vanuatu on July 30, 1980, with the simultaneous departure of the French and British governments.

When I later spoke with Ken Hutton, he told me what had happened: "The bloody Frogs wanted to keep their land, and then you Yanks got involved. Bloody Oliver and his Phoenix Foundation want another Minerva Reef. Lini won't let it happen. Bugger all, my plantation's going to the *kastom* owners—whoever they are. It's going to be a bloody mess to sort out."

Later, while participating in a conversation at the bar, I learned Jimmy Stevens had tried to convince the Vanua'aku Pati and the condominium government to create a confederated Vanuatu with autonomous states. The proposal was rejected.

Back in the States, the next day's news from Santo said the insurgents had blocked the airstrip at Pekoa Field, the airport where Safari Barbie was introduced to John and Karen. Since no planes could land, the government sent ships to Espiritu Santo and evacuated approximately two thousand expats and Santo residents (those who weren't affiliated with Nagriamel). Ken was among them. In late June, John Beasant was appointed as Lini's press secretary and briefed the press daily in Port Vila.

This was the situation when I arrived in Port Vila in late June. My research objective was to return to Malo and collect more blood to identify variants or polymorphisms in their mitochondrial DNA. The hope was to gather information about population migration in the South Pacific, a topic of interest to Luca Cavalli-Sforza, whom I'd worked with at Stanford.

---

2. John Beasant, *The Santo Rebellion: An Imperial Reckoning* (University of Hawaii Press, 1984). On page 96, note that what the reporter transcribes as "yumi" is actually written and pronounced phonetically as "you me," meaning "you and me" or "we" in Bislama.

3. A Bow and Arrow Rebellion," *The New York Times*, June 16, 1980.

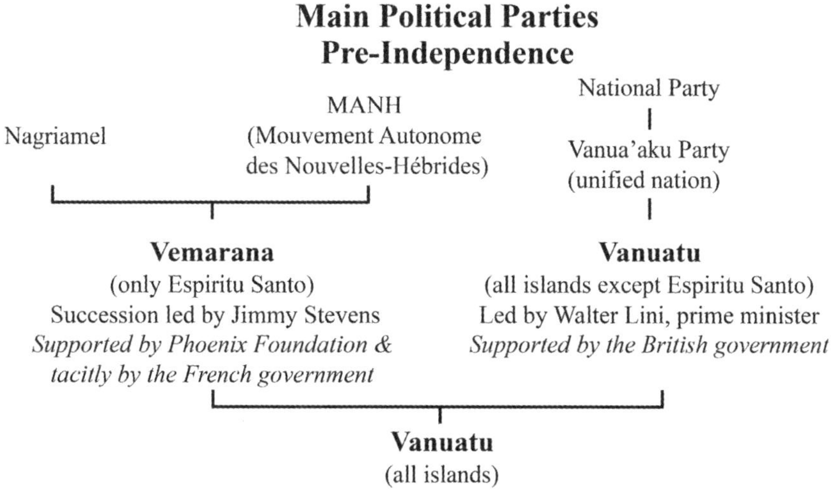

FIGURE 46: Pre-Independence political parties

Port Vila was alive with foreigners, and what better place to meet them than at the bar? I surveyed the room of well-groomed, casually dressed men. A young blond who was alone at the bar caught my eye. I placed myself next to him and waited to order my gin and tonic.

"What brings you to Port Vila?" I asked.

"The revolution. My magazine thought it could be a good story, so they sent me here. What brings you here?"

I gave him the highlights of my years of visiting New Hebrides, my plans for this trip, and my desire to participate in the Independence Day celebrations with my boyfriend, Boyd, who would be arriving closer to Independence Day.

"How are you getting back up north?" he asked.

"I haven't figured that out yet. I was going to fly, but that's hopeless. I guess I'll have to find a boat or ship that's heading toward Malo."

"I have a friend who has chartered a boat. Maybe they can drop you off at Malo."

"That would be great."

"We're heading up tomorrow. While we're on the boat, I'd like to learn more about your work. I can use it for my magazine."

"That's great. Where and when do I meet you?"

The yacht took a full day to reach Malo. Between bouts of seasickness, I got my first taste of what reporting was like. The reporter's summary of what I told him (I considered it to be mundane work) became far more interesting when he described it. He said it was part of his job to create intrigue and write a story that would interest his readers.

When I landed on Malo using the ship's dinghy, I wasn't certain where I was going to stay. I found Vira, who let me use the small hut he'd built for me in 1978. He still had the refrigerator, so I bought kerosene and brought it back to life. I couldn't find anyone willing to ferry me across the Canal into Santo due to the political situation. And, even if I got to Santo, I wasn't sure what the benefit would be. So I opted to remain on Malo and subsist on food from the gardens. I collected and processed blood samples. All was going well. As the weeks passed, I wondered how I would return to Port Vila. There were no boats on Malo that could safely make the crossing to Malekula, where Air Melanesiae still had flights.

Going with the Winds

Just when I'd made the decision to chance passage to Malekula in an open skiff, the Easterlies (trade winds) arrived. The strong winds and rough seas made it impossible to leave the island. Days later, on July 19, when the wind changed back to its usual force and direction, realistic possibilities emerged. While I was waiting to leave Malo, I wrote to a friend at ICR.

July 22, 1980

Dear Barbara,

I absolutely have to do something to kill the time. It is funny how waiting seems to make it impossible to do anything constructive.

Since Saturday, I have been waiting for the ship to take me down to Vila. The cargo ship, Federation Nalkuthan, arrived on Saturday, saying that they were leaving that night, so I didn't take any more blood, just hurriedly processed the samples I had in the refrigerator.

Well, they didn't finish loading the copra and didn't get to Avunatari or to Nanuhu. So Sunday they spent the day anchored off the coast at Nawiambu. Late Sunday afternoon I went on board and spoke with the captain about either getting down to Vila by ship or having him drop me off on Malekula, where I could take the plane down to Vila. He said fine, but they were first going over to South Santo in the morning and would be back around 4:00 p.m. and would pick me up then, along with the hordes of waiting others. The ship is to take the custom dancers and the string band to Vila so they can participate in the Independence ceremonies and festivities. Furthermore, people who worked for the government in Santo are also trying to get down to Vila in order to pick up their paychecks. The problem in Santo is still there. I had forgotten the feeling of helplessness and isolation of being on an island with no way off. It is ironic. I just finished reading John Fowles's The Collector and there is vulnerability that one feels just waiting.[4]

I worry about getting to Vila to meet Boyd, who should arrive just before Independence at the end of the month. If I do not get off the island soon, I may not see him until after Independence. I am scheming ways to try to get to Malekula. Allegedly, the Nalkuthan is the last boat to come here prior to Independence. On the radio, when they report where the ships are, they lie about the whereabouts of the Nalkuthan and say it is in Malekula. I think it may have gone up the west coast of Santo to get copra. I'm sure it will be back, but the waiting is killing me. I can't start taking blood again, since I have no way to carry it on board [the] ship. So, I sit and read, try to get tan, and now, write.

Things have been going well. Tamambo has returned to me. I am not as fluent as I once was, but that is to be expected. So far, there has been no problem in getting blood samples, eighty-four in thirty-one days. I hate it. Each night, when I come home, I tell myself that this is the last time I am going to do this. I am getting too old and I detest taking the blood samples. It's time for someone else to do this. I proved my point. I did it and did a good job. I see no reason to beat this horse to death. I can see returning to get specific answers to questions, but never to do another epidemiological survey. That, I leave to someone else or to a team of people.

It is strange coming back here and going through many of the thoughts and feelings I had when I first came.

Well, the sun's come out and it is 10:00 a.m. Time for me to torture myself lying in the hot sun to become tan and beautiful so I can look good when I arrive in Vila. I'll kill Boyd if he doesn't come, but regardless, I'll look terrific. Nothing like a lot of sun to hide every flaw.

---

4. John Fowles, *The Collector* (Laurel Press, 1981).

Well, I roasted my body at eighty-five degrees for four hours with occasional turning and basting. The boat has come in and is loading the copra and cocoa. So it looks as though I'll make it down. Bye.

I left most of my things on Malo, thinking I'd return after Independence. The blood samples, my typewriter, my tape recorder, my tapes, all the things I needed professionally. I took my Mother Hubbard dresses and a nice batik sundress I'd bought in California. When I boarded, the captain offered me private quarters with a single bed. I declined but did let the captain store my passport and return air ticket, which I'd taken with me for safekeeping.

The ship was filled with people, and it looked like it would be more fun—an adventure and an experience—to sleep on deck. I'd read stories about traveling on cargo ships, and this was my moment. I stayed with a group of Malo women, and we storied on into the night. I got little sleep and was relieved when we arrived in Vila the next morning.

Port Vila was now a completely different bar scene than it had been a few weeks earlier. There were reporters and photographers everywhere, and information was flowing. I found an inexpensive room and scoped out the Independence Day festivities. Ken, who was in exile, helped me secure an invitation to the reception at the British Residency. While I waited for Boyd and Independence to arrive, I went to the crowded bar. There, someone suggested going to the press secretary's daily briefings. Why not? It turned out that being an oddity who was familiar with the geography and culture was an asset. I was the only American in sight, young, single, and female—a rare package. I was also fluent in Bislama and was a great informant about the country's history and culture.

With nothing better to do, I attended Press Secretary John Beasant's daily briefings. Beasant was the perfect choice to be Prime Minister Father Walter Lini's closest adviser. He was from a politically prominent Rhodesian family and had been the press attaché during Rhodesia's transition to Zimbabwe. He had recently worked with Sir Julius Chan, the prime minister of PNG. Beasant was a sensitive, articulate, literary man with effeminate mannerisms and an Etonian accent. I found his daily briefings to be thoughtful and nuanced. From my perspective, he was communicating on at least three levels: First, he presented information on the day's events; second, he subtly inculcated Lini's personal philosophy by frequently using biblical metaphors; and third, he ably diverted the press from crucial issues. Remarkably, Beasant kept the New Hebrides newsworthy. Moreover, he controlled the movement of journalists and exerted autocratic control over the available information.

On July 24, the day after I arrived in Vila, I learned that a joint force of British commandos and French paratroopers had landed in Santo, where they were welcomed to Michael Oliver and Jimmy Stevens's Vemarana with garlands of flowers. Their military mission was to restore the condominium's authority to govern, but their orders were to confine their presence to the town and make no arrests. Journalists had not been allowed to cover the mission and, up until then, no journalists had been to Santo since the succession. The following day, that changed. Under pressure, the Lini government gave permission for a chartered Air Melanesiae flight to leave for Espiritu Santo.

One of my newfound bar buddies was among the reporters who chartered the plane. He asked if I'd go along as an interpreter. You bet! Before I left, I found Ken.

I told him I was going to Santo and asked if he wanted me to get anything. He said, "Bring me the gun I keep in my nightstand in the bedroom. I don't want anybody getting shot with it."

"Okay."

Ken gave me his house keys. Then he said, "You think you'll need a car?"

"It can't hurt," I said, and he gave me his car keys.

On the trip, I learned that the reporters, who were British, Australian, and Czech, represented major newspapers. There was another woman on the plane; she was a photographer for United Press International. Including the pilot, Keith Barlow, and myself, we filled the Islander. On landing at Santo's Pekoa Field, we were met by Georges Cronsteadt, the de facto leader of Vemarana, who cleared us to see Stevens. At the airport, there were no guns and no militia. Since we had been authorized by Cronsteadt, we were good to go.

I found a ride from the airport to Ken's flat, then returned to pick up the reporters. It was obvious that Luganville, the French name for Santo,[5] was controlled by Vemarana—the condominium's military presence was invisible. Driving Ken's packed Honda Civic the thirty-five kilometers to Vanafo (Nagriamel headquarters) on undulating, rutted dirt roads was slow going. One of the reporters said he had experience on unpaved roads and offered to drive. When we switched places, he immediately increased the speed and showed me that going faster gave a smoother ride. The gate

5. Luganville is the name used today.

into Vanafo was draped with *na-mele* (taboo) leaves. I proved my worth when I spoke to the guards in both Bislama and Tamambo, much to their amazement. We waited while the guards inquired about admitting us. Once Stevens gave the okay, they raised the rope and let us enter.

Outside the main building, we met Jimmy Stevens. His entourage of men wore traditional *nambas*. One of the reporters asked if they had any of the carbines they'd stolen from the British police station the night of the coup. The soldiers obliged and returned with five rifles. The reporter had a couple of traditionally clad men pose with both rifles and bows. Once they had their shot of the "Coconut Commandos" in the "Bow-and-Arrow Rebellion," we went inside to talk with Stevens.

FIGURE 47: Nagriamel headquarters, Vanafo, Espiritu Santo

FIGURE 48: Nagriamel fighters on alert

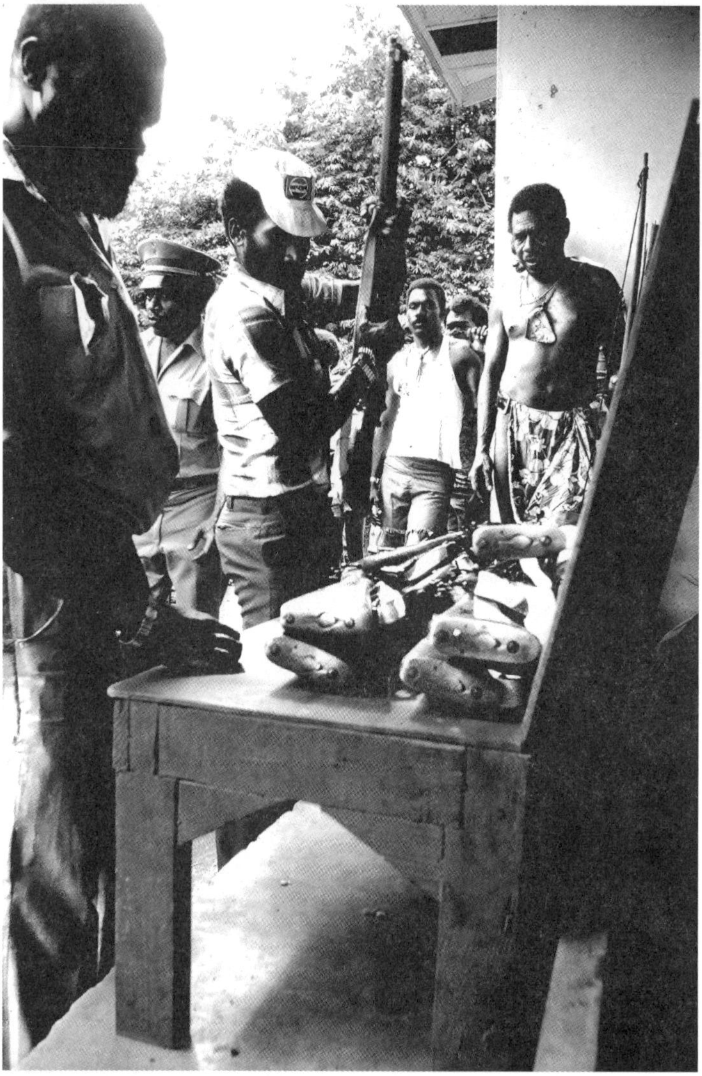

FIGURE 49: Stolen carbines (rifles) from British police station. Followers showing off the guns at the request of the reporters.

FIGURE 50: Vemarana (Nagriamel) custom Land Rover

Jimmy Stevens was an unlikely character to emerge as a local leader. His father was Scottish and his mother was Tongan. He had bright-green eyes, light-tan skin, and a full white beard. I sat directly across from him, less than three feet away. I was so close that my photo of him was published in the *Brisbane Herald*. The reporters asked their questions, and Stevens laced his replies with double entendres.

"Yes, my followers will fight to their last spear," he said. "The Lini government is worth bugger all."

Stevens was very much a figurehead. In addition to Nagriamel, the MANH party on Espiritu also favored succession.[6] This political party was composed primarily of the French and mixed-race New Caledonians who wanted to retain their property. They ultimately backed Jimmy Stevens. His other supporter was the Phoenix Foundation, which provided Vemarana with its constitution, uniforms, passports, and currency—all the trappings of a country. (See Color Plate 8: Jimmy Stevens, president of Nagriamel [1980].)

6. Mouvement Autonome Nouvelles-Hébrides.

FIGURE 51: Reporter taking notes

FIGURE 52: Jimmy Stevens captivating his audience

After the interview, we drove back into town. I was struck by how normal it seemed. *Close Encounters of the Third Kind* was playing at the cinema, and fresh-baked French bread was available in the shops.

"Let's go to Chez Pappy's for dinner," I suggested.[7] I left the journalists there and returned to Ken's, where I left his car. While there, I picked up the pistol and put it into my white canvas book/blood-collecting bag.

Chez Pappy's was the best restaurant in town and the reporters were buying, so why not? While dining on coconut crab (a local delicacy), everyone offered their interpretation of what had happened at Vanafo.

The Australian, who'd been shell-shocked during the Vietnam War, described our visit as "hostile, serious, and dangerous with guns everywhere."

The Czech reporter thought it was interesting and wasn't completely certain about how to report on it.

The Brit said, "There wasn't any violence, not really. It looked colorful more than anything else."

What I had seen was an affable man playing to an audience by using quotable quotes. It became clear that each of the reporters had a very different interpretation of what he had witnessed and about what they were going to write. It bothered me that many of the photographs had been staged. I was shocked at how subjective "objective" was.

At the end of dinner, I mentioned that I had Ken's pistol and wondered whether it was loaded. The Australian offered to check. He took the pistol into the men's room. When he returned, he said, "The gun's not loaded."

"How did you figure that out?" I asked, thinking I might learn something.

"Simple. I pulled the trigger and nothing happened."

"Huh? I guess that's one way to check." I took the gun back.

Over dinner, Michel Portal, the butcher and "Pappy's" son, told us that Santo was frequently being resupplied from Nouméa. *So much for the blockade.*

Someone paid the bill. Somehow, we found a ride to the airport where the pilot had been held, presumably as collateral. When we returned to Vila, I found Ken. I returned his keys, gave him the pistol, and recounted the adventure.

I continued attending Beasant's briefings, and as Independence Day neared, I felt and saw changes in the bars. More military men were ordering drinks. Their uniforms and accents differed. There were troops from France, Britain, Australia, and PNG (Papua New Guinea). The festivities leading up

7. Also known formally as Portal.

to Independence began on Monday, July 29. Boyd had arrived the day before with a sound man from Australia. They filmed and recorded the key events. Unfortunately, I've neither seen nor been able to locate the footage, despite our verbal agreement that I'd have access to it.

Lessons Learned

- Move ahead based on what you yourself can do.
- War zones are interesting.
- Good things *can* happen in bars—stay safe but be optimistic!
- Clever language use can influence on multiple levels.
- Waiting makes it impossible to do anything constructive.
- Take opportunities for adventure, but decline offers that put you above the group you're traveling with.
- The faster you drive on bumpy, undulating roads, the smoother your ride will be, particularly if you're in a hurry. (You can extend that lesson to life in general!)
- Politicians are dramatic performers.
- There is no such thing as objective reporting—there is only the story the reporter wants to see or wants the viewer/reader to see.
- Never test whether a gun is loaded by firing it in a bathroom!
- Verbal agreements count for nothing. Without a written agreement, promises are rarely honored.

Chapter Twenty-Three:

Independence and Insurgents

On the eve of Independence, the British Residency held its final reception. Admission was by engraved invitation. The highlight for me was meeting James Michener. Ken pointed him out, which was good because he didn't match the mental image I'd created. I hadn't expected to see a balding old man with oversized glasses who was about my height. I went over and waited my turn to talk with him.

"I'm Betsy Dickie. I loved *Tales of the South Pacific*. I'm an anthropologist and lived on Malo Island. I've spent a lot of time up in Santo, and I imagine you would still recognize it."

"Nice to meet you," he replied, shaking my hand.

"A quick question. Is it true that Madam Ratard was your prototype for Bloody Mary?" I asked.

"Well, we did spend a lot of time together, and she was a great source and inspiration for me," he replied.

It was time for me to move on as others were waiting. I said, "Thank you. Enjoy the festivities."

At dusk, with a cocktail in hand, I joined the *colons*,[1] expats, and other guests—including the Duke of Kent, members of the press, and government officials—as we watched the Union Jack make its final descent. The flag also descended on Espiritu Santo, which was now formally part of Vanuatu. The official change in government occurred in a far less elegant environment. In front of Parliament House, at midnight—disguised as midday by Film Australia's lights—I watched as Father Walter Lini took the oath of office. The witnesses were primarily Europeans.

---

1. *Colons* are foreign settlers and foreigners who were part of the colonial governments.

The ceremonial changing of the government the next day was extraordinary. The British high commissione, Andrew Stuart,[2] arrived in a Jaguar and stepped out in a gold-braided white military suit wearing a helmet with ostrich feathers sprouting from the top. It was a blast from the colonial past.

His French counterpart, Jean-Jacques Robert, arrived in a Citroën wearing a custom-tailored dark-blue serge suit. They were followed by Prime Minister Lini, who wore a *nambas*, and a troupe of men from Lini's native Pentecost Island who performed a *kastom* dance. The contrasts of the condominium, its governance, and its future were summed up in those few minutes. At noon, I watched as the Vanuatu flag was raised, and by 12:10 p.m., everyone had dispersed. The Vanuatu flag was raised throughout the country, even in Santo, where the infantry raised it on the flagpole in the British Paddock.

The rest of the day was a party. Local bands played a mixture of pidgin songs, which are an amalgam of gospel, bluegrass, and country.[3] Boyd and I danced, drank, and had a great time. Revolutions are interesting, and independence parties are *fun*.

It was a good and memorable day.[4]

Blocked

The next day, Boyd's other sponsor,[5] W. Scott Malone, arrived from New York. He wanted to go to Santo to meet Jimmy Stevens, but the blockade was still in force. On August 2, we flew to Norsup, at the northern end of Malekula, hoping we'd find a boat to take us to Santo.

At the airport, I talked to the driver of a truck taxi and explained our situation.

"Get in my taxi," he said. "I can take you to a boat."

---

2. "Andrew Stuart–Obituary," *Daily Telegraph*, February 7, 2014, and "Andrew Stuart: Diplomat Who Fought 'Coconut War' for Vanuatu Independence," *Sydney Morning Herald*, February 16, 2014.
3. "Dollar Nomore, Makem Me Loosem Home" by Fred Maedola of the Solomon Islands is a good example. See "Dollar Nomore, Makem Me Loosem Home," posted June 17, 2021, by Fred Maedola, YouTube, https://www.youtube.com/watch?v=0KYJ7NSUo0M.
4. A documentary video of Vanuatu Independence Day, *YUmi Winnem Freedom*, is available here: "Vanuatu History Documentary (YUmi Winnem Freedom)," posted June 24, 2015, by Yumi TokTok Stret, YouTube, https://www.youtube.com/watch?v=m8VfxQpvEKA&t=3s.
5. Boyd had connected with Malone, who was also a freelancer and had been hired to film the Independence ceremonies. Since Boyd's travel was paid for by the University of Illinois, it was cost-effective for Malone to hire him as the cameraperson. Malone also hired a sound technician from Australia.

"How much?" I asked.

"Sixty dollars, Australian," he said.

I looked at Scott and asked, "Should I agree to sixty dollars?"

He nodded.

"Okay," I told the driver. We loaded ourselves and our equipment into the truck.

He drove us an hour and a half farther north, where we had to pass through a Nagriamel checkpoint.[6] Again, my ability to speak Bislama and Tamambo proved useful. We were cleared and a large boat appeared an hour later to take us to Luganville. As luck would have it, I knew Michel, the owner of the boat, from my fieldwork days. He and Scott agreed to a price and we were off, crossing the channel and bouncing our way first past Malo and then Aore, up the unsheltered eastern side of these islands. When we arrived in Santo, the same Georges Cronsteadt who'd met my chartered flight with the reporters was waiting at the dock and gave us clearance to visit Jimmy Stevens.

Scott[7] was freelancing for a television channel. He rented a car, and the following morning, we traveled to Vanafo. Upon our arrival, we were officially welcomed by Nagriamel's custom chiefs and by the Vanafo tribe. The men, mostly in native dress, stood in three lines, shoulder to shoulder, facing the Nagriamel Office Headquarters. To the tune of "My Country 'Tis of Thee" (which, ironically, is the same tune as "God Save the Queen/King," the anthem of the colonial power they disliked), they sang the official Nagriamel "Welcome Song." Its lyrics were in English. This was a made-for-television moment, which Scott had the documentary duo (cinematographer and soundman) record. It was very different from the reception the traditional pen, pencil, and photo reporters had received.

We had been waiting about twenty minutes when the showman extraordinaire staged his entrance. Since I had caught Stevens's act a week earlier, I was interested to see what he would choose from his repertoire for the Americans. No longer was this the comic opera "Revolutionetta" to be armed with coconut toddies. Now he was asking for guns, ammunition, and cash.

---

6. There were isolated Nagriamel outposts on both Malekula and Malo even though both islands were officially part of Vanuatu (as, technically, was Espiritu Santo). Getting into Santo still required Nagriamel approval.

7. His account appears on his website: W. Scott Malone, "W. Scott Malone: Dilemmas of War—The Media," BlackNET Intelligence Channel, May 5, 2012, https://blackboxintelcenter.blogspot.com/2012/05/w-scott-malone-dilemmmas-of-war-media.html.

He appealed for sophisticated automatic and semiautomatic weapons—nothing like the single-action carbines he had taken from the British police station. He stressed the need for substantial American backing and arms so he could continue the fight for "free will." At all times, Stevens was affable and accommodating. He even offered us Vemarana passports. Boyd took one but I declined, remembering my signed statement from 1975 when I had agreed to neither participate in politics nor express a political opinion.

Stevens was politic and sensitive to the desires of his audience, allowing them to satisfy their needs through him. For the French, Stevens could produce a block of voters. For Vemarana, he was the New Hebridean figurehead legitimizing the revolt. For the press, he was a colorful figure and made good copy. For Michael Oliver, he was a sympathetic Melanesian entrepreneur. And for Scott Malone, the American television producer, he was a political pawn caught up in international intrigue.

FIGURE 53: Filming the rebel army

FIGURE 54: Jimmy Stevens. Name badge: President, Nagriamel, Jimmy "Moli" Stevens." Posing for film crew.

FIGURE 55: Jimmy Stevens interview

The next day, I looked for a way to return to Malo. There was none as the blockade was still in force. We were offered a "ride" back to Vila in a Royal Air Force Hercules C-130 military transport plane. I took the flight—my first and only time in a military plane. I left Vanuatu within a week and never made it back to Malo on that trip, but I was able to take a casualty loss on my taxes.

I spent my next to last day in Vanuatu having lunch with a young Malo man, Nikenike Vurobaravu.[8] I'd taken blood samples from his father, mother, and siblings. We had first connected in Santo in 1976; we'd enjoyed talking and sharing a meal. After the first meal, we had met for lunch a couple of times during my previous trips, usually in Vila. Nikenike was one of the few people in New Hebrides with a university degree—a BA in political science from the University of the South Pacific in Suva, Fiji.

During our last meal in Vila, post-Independence, he described his present situation: "I'm an officer in the Department of Foreign Affairs. I am honored to represent Vanuatu at the United Nations."

"That's great! What's it like? How do you like Manhattan?" I asked.

"New York is another world, and the Papua New Guinea delegation gives the best parties."

I was thrilled that a Malo man from a remote village had done so well and risen so high so quickly. After a distinguished career in foreign affairs, he was elected president of Vanuatu in 2022.

Soon after Independence, Vanuatu signed a formal defense treaty with PNG. Vanuatu's prime minister, Father Walter Lini, then requested that the PNG government send troops to relieve the joint French and British forces in Luganville. The Vanuatu government offered amnesty to the rebels, but they rejected it. Lini then directed the PNG troops to arrest those rebels on Espiritu Santo who would not surrender. Jimmy Stevens was arrested. The two dozen French and mixed-race provocateurs behind the rebellion either were deported or escaped to New Caledonia. The final casualty report for the ninety-day rebellion included one death and one injury, both accidental.

---

8. He was from the village of Atariboe at the southern tip of Malo Island. The village was so far from Avunatari that I only visited it two or three times.

A Matter of Ownership

What happened with the rebellion, and why? Here are my observations and theory. The pivotal issue was land ownership. The French and Nagriamel wanted Vanuatu to be a confederation, with each state/island determining its own land policies. Under this scenario, the French kept their plantations and Oliver got a capitalist utopia.

British-backed Lini's vision was to return foreign-owned land to its *kastom* owners. At the time, I wondered whether Beasant and Lini had deliberately handled the coup with a light touch. They told government supporters to evacuate from Luganville, leaving only those who supported the rebels.

Lini asked the French and British to send the joint military forces to Luganville. The troops ensured that the Vanuatu flag was raised on Independence Day, but they were under orders to take no action against the rebels and to stay on the grounds of the French and British Residencies. This act ensured that Vanuatu, as a united country, would continue to receive aid from both the French and British governments.

After Independence, the new Vanuatu government offered the rebels amnesty. They rejected the offer, thinking a separate state might still be possible. However, the Lini government rejected that idea and gave the rebels the choice of either being tried as traitors or being deported. The New Caledonians chose deportation. The French, who had tried to have it both ways, were seemingly outmaneuvered by Lini—an impressive accomplishment.

Using the PNG Army to take back Luganville was intended to show the world that Melanesia was to be run by Melanesians. At the time, there was talk that this action sent a warning to the colonial powers in New Caledonia (France) and Western Papua (Indonesia), especially since representatives of the independence movements from both New Caledonia (the Kanak Party) and West Papua (West Papua Party) were invited to the Vanuatu Independence celebrations, while representatives from Indonesia were not invited.

Resolving the rebellion in Vanuatu showed that an ex-colony could solve its internal problems without turning to European governments for help. Papua New Guinea also used this occasion to establish itself as the dominant island nation in the Southwestern Pacific.

And Jimmy Stevens? He surrendered and was tried for treason[9] but gained international notoriety. He became nationally recognized and was taken seriously. The Lini government even implemented many of his original land policies. As the *kastom* landowners looked for credibility and a new identity in independence, Stevens became the opposition's de facto leader years before the rebellion. At the time, I wondered whether Stevens would someday be president or prime minister of Vanuatu. If so, he would have been the only man to lead both Vemarana and Vanuatu, all the nations of the New Hebrides.[10]

When I returned to Vanuatu in 1982, I sat next to the chief justice of Vanuatu's supreme court on the plane. I shared my theory that Lini and Beasant had orchestrated the aftermath of the Santo coup to a united Vanuatu's advantage.

He smiled and nodded. "You might be more right than you know."[11]

Lessons Learned

- Parties are opportunities to introduce yourself to important people, and those memories last forever.
- People's behavior and requests vary depending on the context and opportunity.
- Hitching a ride on a military transport is an interesting and unique opportunity.
- Losses in a war zone can be used for income tax write-offs.

---

9. It all came to a sudden end when a truck carrying Jimmy Stevens's son careened through a roadblock in August 1980 and soldiers opened fire on it, killing the young man and bringing about Jimmy Stevens's surrender.
10. He was brought to trial and sentenced to fourteen years in jail (of which he served eleven). The trial revealed that the French government had provided him with secret support. He was arrested in November 1980 and released in 1991.
11. See "More Tales from the South Pacific: Independence New Hebridean Style," a contemporaneous account written in 1981, on www.ElizabethReedAden.com (under tab *Mud, Microbes, and Medicine*).

Chapter Twenty-Four:

Detours off the Express Lane

When I left Vanuatu in 1980, I took some detours on the return flights. First was a visit to Australia to catch up with Ted and Heather, who had moved to Brisbane. I stayed a few days, and they introduced me to koalas and kangaroos up close. My next stop was a couple of days in Fiji, where I had planned to sightsee but instead was trapped by torrential rain. Stuck in my hotel room, I wrote my postdoctoral fellowship application to the National Institute of Allergy and Infectious Disease (NIAID). If the grant was funded, it would support me for three years at Stanford with Luca Cavalli-Sforza.[1] I assumed that I would work out the details with the University of Illinois.

On my way back to Illinois to teach the fall semester, Boyd and I spent time together in California at my parents' summer cabin. Our relationship ended that summer when he opted not to meet my parents. His decision clarified my thinking about whether this would lead to marriage—nope. He returned to Boston soon thereafter, and we've remained friends. I flew back to the miles of endless cornfields in Illinois and taught the fall semester.

I was constantly exploring ways to make teaching fun. At first, it took me four hours to prepare a fifty-minute lecture. I would sit in the local diner with a breakfast of eggs over easy, hashbrowns, and endless cups of black coffee. I'd pull out my books and articles and write out the points for my human biology lecture. I was usually finished by the time the lunch crowd arrived. By midsemester, I'd figured out that my days could be more efficient if I posed questions and let the students spend class time figuring out the

---

1. I knew a two-year leave of absence was possible from the University of Illinois. For the third year, I'd cross that bridge when I came to it.

answers. Teaching was exhausting, although the bright, engaged students made it rewarding.

I returned to Stanford for the spring and summer semesters of 1981. It was there I learned that my postdoctoral fellowship would be funded. Cavalli-Sforza was impressed by the near-perfect score my application received. The Individual Postdoctoral Fellowship was awarded to me—not to Stanford. This gave me more academic freedom, but it didn't spare me from encroachments on my personal space; unwanted advances abounded.

On one occasion, when we were in his office with the door closed, Cavalli-Sforza said, "I'm sure you slept with Blumberg, so why not sleep with me?"

"I *never* slept with Blumberg," I told him. "And I'm not going to sleep with you."

"No? Well, why not?" he persisted.

At that time, it was common for famous professors to make overt sexual advances to women students and staff members. The medical school was a world of male professorial entitlement. Faculty members acted as if they were untouchable, and sexual vibes were everywhere. It was palpable and uncomfortable.

I needed advice and context, so I bought Machiavelli's *The Prince*. It became a useful guidebook. Stanford was rich and the institution recruited world-famous faculty. These academic princes had their empires, acolytes, and money. They could get away with anything, and they tried everything. If you were an attractive woman, you were fair game. It was a game I didn't play.

As part of my research, I planned to live in Taiwan starting in 1982. I would do a project similar to my doctoral work—I'd compare and contrast both the epidemiology and natural history of hepatitis B in Taiwan and Vanuatu. Similar to Vanuatu, hepatitis B was endemic in Taiwan, and the frequency of the virus and chronic carriers was very high. However, the incidence of liver cancer in Taiwan was high, whereas it was nonexistent in Vanuatu.[2] I wanted to investigate whether there were other differences, since Taiwanese children did not become chronic carriers until after the age of two.

I wondered whether this difference in timing was critical for developing liver cancer. I wanted to understand if and why seroconversion was so much

2. Thompson was still alive in 1980, so I knew of no cases of primary hepatocellular carcinoma in the New Hebrides.

later in Taiwan, and I expected to study Taiwanese child-rearing culture and behaviors. I audited Mandarin language classes so that I would be able to fit in better—but something else was about to turn my world upside down.

I met Mel.

Meeting Mel

Before my professorship at Illinois began in the spring semester in 1980, Luca Cavalli-Sforza's lab sponsored me at Stanford from October to December. My friends at Penn offered to help me start a social life in California. Kay came to the rescue.

"I have this old boyfriend, Brian, who is a management consultant," she told me. "He's doing a project at Stanford and, even better, has an expense account. You'll love him. I'll introduce you."

That seemed fine. I went out with Brian, who soon had a birthday party. It didn't take me long to notice that Brian had a very cute friend, Mel, an engineer and private pilot who was at the party with his girlfriend Jane. Mel and I hit it off right away and had a good time chatting and laughing. Later, the four of us arranged to go skiing and stay at my parents' place in Olympic Valley for the weekend while I finished the index for my dissertation—the final procedural piece.

When we got to the cabin, which was within easy walking distance of the ski slopes, it all seemed fine—very nice, very civilized. But I discovered Brian was cloying. It was as if he had to be in my presence all the time, which I didn't like at all. Mel was different. Mel and I could ski, but Jane and Brian couldn't. Mel was easygoing and relaxed; Brian was not.

It all came to a head—although not a dramatic one—when I left for Illinois. I didn't tell Brian that I no longer wanted to see him, but I suggested that he ought to look up my friend Carol in Los Angeles, since he was based there. Soon thereafter, I heard that they were an item and madly in love.

Well, I thought, *this is interesting.*

So, I called her. "Hi, Carol. How's it going?" I asked innocently.

"Oh, Betsy. I have a confession to make . . . Brian and I have been dating."

"Carol, that's great!"

"I didn't want to tell you, because I thought you still liked him," Carol said.

"I like him, but I'm not interested in dating him. I'm so glad that you are—that's what I'd hoped for when I provided the introduction."

"I'm so relieved. We're actually quite serious and are talking about getting married."

"Carol, that's fantastic. Tell you what. To assuage your guilt, perhaps, just possibly, you might tell Brian's friend Mel that I'm available?"

Guilt works fast—Carol called Mel. A couple of reminders from Carol resulted in Mel calling me. Super! We made plans to meet now that I was at Stanford full-time, having been awarded the NIAID Individual Postdoctoral Fellowship and a leave of absence from Illinois. I'd wanted to know him better since the ski trip ten months earlier. We were lucky that Brian and Carol made it easy to connect.

However, Mel's father died on the day of our first planned date, so we rescheduled. When we did go out, I watched with pleasure when he limited himself to a single glass of wine. For the next three months, he suffered through listening to Mandarin tapes in the car. I practiced every moment I could because I was planning to leave for Taiwan at the end of the semester.

Mel often wrote about our new relationship in his diary:

November 16, 1981

Betsy is exciting, good, and an eager hostess. ∞ Wine, perfect Italian appetizers, "green" lasagna, and carrots al dente.[3] *We did not talk enough but felt close and hugged goodbye.*

November 23, 1981

Saturday, it rained all day, but the day was saved by Betsy, that bundle of energy and excitement. She wanted a windsurfing lesson, so we missed each other for one hour at Redwood Shores. She is fundamentally coordinated, and will learn easily, and I am impressed by anybody who will want to get a lesson in the rain! She spoke of her problem: high energy level and finding people who can keep up. Don't look at me for that . . .

She also discussed her career. New Hebrides circa 1975, 1976, and previous studies 1973. I was touched by discussion of [her] New Hebrides grieving process. But I wonder: How sensitive is Betsy to my situation with respect to my father? The issue has never been acknowledged between us, and I'd like that. High-energy façade can "cover" a lot.

3. Mel, an electrical engineer, used the infinity symbol in his notes, so I've retained it.

She's very comfortable to be with. Saturday the winds were blowing twenty to forty-one [knots] so we scrubbed the windsurfing race. Made omelets (outdid myself again) [and] went riding with Shar, Betsy's horse. Worked in the evening, she read Aviation History of Medicine,[4] *and then we danced ourselves to bed.*

January 2, 1982

We went to Mike's overgrown trailer house on Skyline near Gist. And we spent three hours in the hot tub, including dinner in the tub, a great luxury! Then we sang songs for hours on end, and Betsy knew the words to most all of them! She's great! I really like her!

January 5, 1982

Betsy's birthday party was well received by all. Surprise was total, as she came with Ed [a flying buddy and instructor] after flying. My fun games included making food under her face, such as bacon, onions, and coleslaw, and designing a crossword birthday cake.

I was ambivalent about the relationship because I had plans to live in Taiwan for a year. However, I was also occasionally seeing the psychiatrist my mother had sent me to when I was ten, soon after she married Sal. Emmy Sylvester came to the rescue. I followed her suggestion.

"If you don't want me to go to Taiwan, make me a better offer," I told him. He did.

Mel's diary elaborates on the events that occurred four months after our first date:

March 17, 1982

1. *Everything happens at once and very happy. Betsy and I became engaged last night.*
2. *Saturday and Sunday we go ring shopping. Monday likewise. Spot her ring at Antoinette's.*
3. *Tuesday [morning] I call, purchase ring, and have it fitted.*

---

4. Douglas H. Robinson, *The Dangerous Sky: A History of Aviation Medicine* (University of Washington Press, 1974).

4. *I take an early train to the City [San Francisco], pick up ring, go to her apartment to put her bottle of Dom Pérignon into the refrigerator.*
5. *I take a cab back to the Southern Pacific train station and wait for Betsy.*
6. *She is an hour late. I freeze to death, fear her dead, and even try to call Dr. Sylvester to see if she showed up.*
7. *We go to the Albatross Saloon for quick dinner (calamari fritti).*
8. *Attend splendid performance of Evita at the Curren Theater.*
9. *Go to the Fairmont Hotel Crown Room for a drink. I ask her to marry me. She asks if it is for real.*[5]
10. *I repeat several times, offer knees, get down on bended knee.*
11. *She keeps me in suspense as I drink her mai tai, since it has a straw. Too nervous to pick up my own drink, margarita.*
12. *She says, "Yes . . . I like the cable cars." Keeps teasing. Finally she says, "Yes, I'll marry you."*
13. *Great relief! We kiss, hug, & cuddle.*
14. *I produce bottle-top pull, too small for her finger. She thought I was going to produce a ring. She looks relieved but disappointed.*
15. *I produce a camera followed by my ring box.*
16. *She can't believe it is for real—total shock. Betsy nearly trips over her dangling jaw.*
17. *After procrastinating, she discovers that it is the ring she wanted, is flabbergasted that it was possible to get it.*
18. *Betsy is all over me to explain. She remains shocked, pleased; we say loving things.*
19. *I drive her home. She does not know which way we are going.*
20. *At home, she cannot find her bottle [of] champagne. I find it in the fridge, as if that is it. [She is] shocked again but starts to produce plausible explanations.*
21. *Other things follow that are not to be written down. The champagne was the best of my life, but anything would have been fine, too. I keep Betsy.*

---

5. Even though we'd gone ring shopping together, I still didn't believe he would propose.

Turns out that I had inadvertently taken Mom's advice:

April 27, 1975—From Mom

I just hope that you will carve out for yourself a life that will be enjoyable and not one of drudgery and meanness. And it seems to me that those I know in immunology are more narcissistic than those in the humanities fields—but then it is really impossible to generalize—as people are so different. Just be sure the man you get involved with is really good to people, likes people, respects you, is good to you, likes women, and had a good relationship with his natural mother and father. You have had enough problems, so don't compound that with the person who also has had enough problems or you will end up with an atomic bomb!

Or as my friend Mary told me, "He was just the person we all hoped you would marry but whom you didn't have the good sense to find!"

Mary was right. Mel was reasonable, smart, and unselfish. He was nothing like the other men I had dated. He was a completely normal person, more like my father, Murray, than anyone else had been. I found him to be perfect for me. He was both easygoing and driven, but in a way that didn't need a lot of ego support. He was more than willing to meet me halfway, and he had his own life. Mel had his private pilot's license and belonged to a flying club in Palo Alto where he could rent small planes. He was an electrical engineer and co-founder of a startup company. When he was around, he was flexible and would give me all the space I needed. It was a straightforward, comfortable relationship from the start.

Seven months after we started dating, we were married.

I never went to Taiwan. I took the better long-term offer. But I also arranged to undertake a project that explored whether genetics contributed to susceptibility to becoming a chronic carrier of hepatitis B; that was a reason to return to Vanuatu, and it would become part of our honeymoon. (See Color Plate 9: Mel and Betsy's official wedding photo [1982].)

Lessons Learned

- Following friends' suggestions can lead to good things.
- Academic status differences make some powerful professors feel entitled to make sexual advances to women, especially younger women students.
- Don't be afraid to let go and let others have a turn—the outcome might be much better than expected, for everyone!
- Sometimes it is good to follow your psychiatrist's advice—and your mother's.
- Good things can come in small packages.
- It is always good to have a bottle of champagne available . . . just in case.

COLOR PLATE 1: Champagne Beach and Hat Island (Espiritu Santo). One of the finest beaches in the South Pacific.

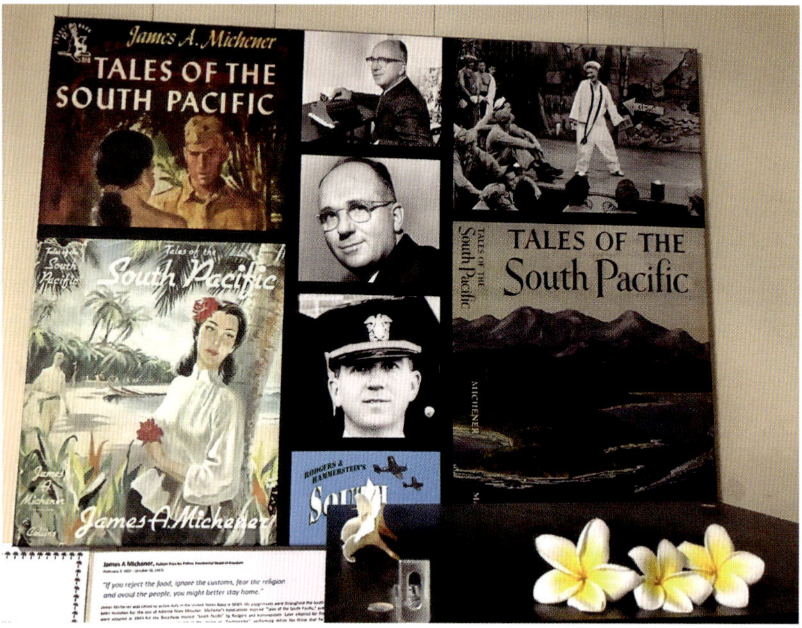

COLOR PLATE 2: *Tales of the South Pacific* by James A. Michener. Photo of display, used with permission of the South Pacific War Museum (Espiritu Santo).

COLOR PLATE 3: Antas's grave. Antas was Jole's grandfather and Sumbue's great-grandfather; he was the first convert on Malo by Presbyterian missionary John Landels. Antas died September 15, 1897.

COLOR PLATE 4: Small Nambas hunting with bamboo bow and arrow. Grant Dixon photography

COLOR PLATE 5: Dining room and office. Main area for eating, working, and conversing. Mosquito netting covers the windows. An Aladdin kerosene lamp provides light.

COLOR PLATE 6: Traditional Malo celebration attire. A ceremony to inaugurate the new well, attended by the British district agent, was celebrated by custom dancing.

COLOR PLATE 7: Boy pumping well water. Bicycle pedals with chain and sprocket are used to pump water from the freshwater lens.

COLOR PLATE 8: Jimmy Stevens, president of Nagriamel (1980). Stevens led the Coconut Revolution for succession of Espiritu Santo from Vanuatu.

COLOR PLATE 9: Mel and Betsy's official wedding photo (1982)

COLOR PLATE 10: Mel with Vira and Voligo's Family (1982)

COLOR PLATE 11: Sumbue and Rosemary Antas (Geneva, 2019). Sumbue is the ambassador and permanent representative of Vanuatu to the United Nations Office.

COLOR PLATE 12: Outdoor market in Santo (Luganville, 2019). A wide array of fresh, affordable garden crops are offered.

COLOR PLATE 13: Fruit bats (flying foxes). A local delicacy. These bats were offered for sale at the market in Santo (Luganville).

COLOR PLATE 14: Tavuti, Betsy, and Alan. Sumbue's son (Tavuti) and youngest brother (Alan) in the kitchen at Avnaro, where we discussed my dissertation findings.

COLOR PLATE 15: Nanda Blue Hole. A sinkhole / underground cave in limestone (coral) filled with seawater. During my visit, I was the only person and swam to the open sea.

COLOR PLATE 16: Marshall and Argyri's wedding (Hydra, Greece, 2023)

Chapter Twenty-Five:

White Cells, Red Cells, and the Blues

My next hepatitis project was to discover if genes determined whether a person became a carrier when exposed to the virus. This research required drawing venous blood from entire families where at least one parent was HBV positive. From these samples, I was to isolate the leukocytes—the white blood cells—for analysis of the human leukocyte antigen (HLA) genes that can affect a person's response to an infectious disease. I knew nothing about the HLA system and nothing about white cells. I couldn't rely on Scott to help me this time—she had died of cancer—so I was on my own.

Fortunately, another woman came to my rescue. Rose Payne was a pioneer in HLA typing and function. I spent a month in her lab testing blood from patients who were candidates for solid organ transplantation. Payne taught me how to prepare the sample and evaluate the result. That part was relatively easy. However, her lab had electricity, ultracentrifuges, and liquid nitrogen for freezing samples. The obvious question was: How would I do this on Malo, where none of this existed?

There was only one place to go to look for answers—*Index Medicus*, a series of bound volumes that you searched by hand. (The PubMed search engine for accessing the MEDLINE databases of articles in the life sciences hadn't been established yet.) It took forever, and you often had to try many different keywords before finding the one(s) that listed relevant articles.

Back to Basics

I was looking for basic and "primitive" methods for preparing and testing blood samples. Surely, techniques had existed to separate white cells from blood in the old days before ultracentrifuges. My search began in the library stacks where the older *Index Medicus* volumes were shelved. I grabbed one decade at a time, took it to a library table, and searched. I found a protocol published in the 1920s that required minimal centrifugation and no fancy equipment. This method used reagents that included gum arabic and iron filings. I showed this recipe to Payne, who speculated that the gum arabic could cause one cell type to clump, which would create a gradient. Intrigued, she thought the macrophages[1] would devour the iron filings and drop to the bottom of the test tube. The procedure used a simple, basic centrifuge. She thought the protocol just might work.

A good first step. But, in Vanuatu, I would have to get even more basic and find a way to centrifuge the samples. The science equipment catalogs included a ridiculous-looking propeller centrifuge that could spin two test tubes. The instrument was cumbersome and long. Fortunately, salad spinners had just been introduced in department stores. I saw one and thought, *Hmm. That might just work.* I bought a plastic salad spinner. The orange lid had a handle that spun a perforated basket inside a semitranslucent bowl. The number of revolutions per minute was determined by how quickly or slowly I turned the handle. Five dollars later, the portable, lightweight centrifuge was mine.

I figured out how many revolutions per minute I could get out of the salad spinner, and it would work. Then I grabbed some test tubes from the lab and brought my spinner to the engineering department at Stanford. "Can you make something that holds six test tubes?" I asked the engineers. "Please convert my salad spinner into a field centrifuge so I can collect the white cells."

They made a Plexiglas holder for six test tubes and drilled holes in the spinner's basket so everything fit perfectly. I had my field centrifuge.

Next, I tried the dubious recipe in Payne's lab and the protocol actually worked! The remaining challenge was figuring out how to freeze and store the precious white cells at minus twenty degrees centigrade. I needed either

---

1. Macrophages are a type of white blood cell that engulf and digest foreign particles like viruses and other pathogens, including cancer cells, microbes, and cellular debris. They are an important part of the immune system.

dry ice or liquid nitrogen. I went back to the equipment catalog and found a compact dry ice maker that required no electricity, just access to a large tank of liquid carbon dioxide (CO_2).

No problem—there was a soft drink company in Port Vila. My plan was simple. I would go to the bottling plant, purchase tanks of carbon dioxide, and then transport these huge, heavy tanks to Malo with my dry ice–making machine. What could go wrong?

I was set. I shipped everything off to Vanuatu. When Mel and I arrived, we spent five days of our honeymoon flying between Santo and Port Vila trying to clear my equipment through customs without paying a fortune for its release. The problem? Pounds versus kilograms. My shipment had been weighed in pounds, but duty was assessed in kilograms. This meant the cost was going to be 2.2 times higher than anticipated. But sort it out we did.

Our next goal was to buy a tank of CO_2 for my dry ice maker. This led to another challenge I hadn't counted on: The American fittings on my machine were not compatible with the British tooling on the CO_2 tanks. We asked around and found someone who could make an interface. Thus, a different type of Anglo-American connection was made. The soft drink plant let me borrow three large tanks, so we packed up our supplies, found a boat, and sent everything off to Santo. The next day, we flew to Espiritu Santo and spent our "down" day swimming at Champagne Beach.

We brought all the equipment to Malo by speedboat. Vira and his sons had built us a high-end hut with a concrete floor, and, at the last minute, they added an outhouse. While we were on Malo, Mel got to attend a wedding and swim in my reef.

"I didn't know that moray eels could slither across the top of the reef," he said the first time he walked out on it.

"What are you talking about? There aren't any moray eels here. The only moray eel is way over there," I said, pointing to an area about two hundred yards away.

"You have twenty-four hundred vision," he countered, "and are so near-sighted you can't see anything."

"True. That's why I don't touch anything when I swim."

He later told me that he'd been chased by a sea snake while swimming there, and that the ocean was filled with them. My original fears turned out to be real! Since seeing is believing, not seeing allowed me to live in ignorant bliss.

FIGURE 56: Our $39 honeymoon hut. Located at Malahatou, Vira and Voligo's oceanfront property.

FIGURE 57: Moray eel reef. The spot where Mel watched moray eels slither over the exposed reef. [See also Color Plate 10: Mel with Vira and Voligo's family [1982].]

We set up the equipment on Malo then left to finish our honeymoon with a visit to Mount Yasur, a volcano on the island of Tanna. Yasur is the most accessible continuously active volcano in the world. A guide drove us to the base, and we followed him up to the rim of the volcano. It was dusk and we were the only people on the rim staring down into the caldera of molten lava. There was a ledge a hundred feet lower, and we asked if we could go there to get an even better view. The guide said no. It was good advice because a few minutes later, a red-hot pumice bomb landed in the middle of that ledge and exploded. Each of the numerous eruptions sounded as though we were inside the engine of a 747 revving for takeoff. It was terrifying, magnificent, and unforgettable.

Before Mel left, we had a last supper in the Pandanus restaurant in Port Vila. It was a bittersweet dinner because this was the restaurant Ted and Heather created. It had been their dream and the reason they came to New Hebrides, but it drove them into bankruptcy. Mel then continued on his solo honeymoon back to California and I returned to Malo.

Exposure

This time, I needed to draw at least ten cubic centimeters of blood (not just a few drops), and that meant using vacutainers. I hadn't practiced for a few years, but the technique does come back—not quite like riding a bicycle, but close. The men's veins were especially tough, like trying to pierce plastic tubing. I had to push hard to get the needle into the vein.

During this visit, I began to feel like a used car salesperson. I was asking people to do things that I wasn't sure had the value I'd once thought. I didn't feel great taking the large-volume blood samples. The first-generation hepatitis B vaccine had just been developed, but I didn't know whether it would be good or bad for the islanders to get it. I felt as if I was constantly having to plaster a smile on my face and say "Hi! I want to take your blood today." Afterward, I'd give them a red-colored vitamin pill, which contained iron, to help overcome any anxiety about the amount of blood I'd taken.

I thought that if I understood how the virus was transmitted, it would be possible to develop a culturally relevant, noninvasive approach to prevent transmission. For Malo Island, the simplest solution would have been introducing a pesticide that eliminated head lice—everyone had them, and lice were the easiest piece of the infection equation to remove. In addition, if we knew whether susceptibility to chronic infection had a genetic basis, then it would be possible to vaccinate those people who were at risk and prevent liver cancer.

Back in my makeshift lab/hut, which held the same kerosene refrigerator I'd used in the past, I separated the red cells and the plasma in the vacutainers as Scott had taught me. I turned the tube upside down and let the clot attach to the stopper. A day later, I gently opened the container and transferred the plasma into another tube. I then returned the clot to the vacutainer and stored it. Next, I separated the leukocytes with my alchemist's protocol. I saved everything, and hundreds of samples later, I had liters of fluid to manage—most of it positive for HBV. I was handling a much greater volume of blood, and it was daunting. I spent every day collecting blood and every night processing it. My time on the island was limited by how long the dry ice would last.

The day I made my first dry ice was unforgettable. This was the one step I'd never practiced in the lab. My machine connected the CO_2 cylinder to a mold that would create a block of dry ice. The cylinder contained CO_2 in both its liquid and gas phases. With Vira's help, I lowered the cylinder so the liquid CO_2 flowed into the fitting. The difference in pressure between the liquid phase and the outside world would create dry ice molded into a block.

We did all the preliminary steps. Then I turned the valve with a great deal of trepidation. The sound the tank emitted was terrifying, and the cylinder vibrated unexpectedly. Vira actually left the hut; he was afraid the tank might explode. I had similar thoughts but stayed and watched until, miraculously, a block of white material emerged. I turned off the valve, picked up the precious lump, put it into a Styrofoam container, and added my prized leukocyte samples. This ritual continued for a week, until I depleted my CO_2 supply. During my last night on Malo, I sealed the test tubes containing the white cell samples, hoping the seal was airtight. Vira offered to wrap the test tubes containing the red cells with Parafilm, a waxy, waterproof, self-sealing material.

The next morning, I took everything over to Santo and left the cylinders with the local distributor. By early afternoon, I had deposited the leukocytes in the freezer at the hospital in Port Vila, where they were stored at a temperature of minus seventy degrees Fahrenheit. I arranged for the samples to be repacked in dry ice and delivered to me at the airport a half hour before my flight left for Australia.

That night, in my hotel room, I was finalizing what I needed to do and checked the red cells. To my horror, the Parafilm-wrapped tubes were leaking and bloody. The seal had not held. I unwrapped each tube in the sink, secured the stopper, washed the Parafilm, and rewrapped the tube. All of this with my bare hands. I looked at my fingers and saw minor openings in the skin—hangnails. Not good. The only available disinfectant was soap and water. After seven years of working with hepatitis B, I'd finally been seriously exposed.

The next day, I went to the airport with all my things and waited for the leukocyte delivery. I called the hospital and was told they were leaving. Leaving? Uh, so was my plane. It was boarding, but I wasn't going anywhere without my samples. I had to get to Australia to send the samples onward to South Africa for testing, and now I also had to quickly return to the US to get the HBV vaccine. There was only one thing to do. I walked onto the tarmac and stood in front of the airplane.

"You've got to move! You can't stand there," the pilot of the Air Pacific 727 shouted down at me from the cockpit. He was a middle-aged Australian and wore his pilot-in-command regalia.

"I can't. I have blood samples that have to be on this flight, and they should arrive soon."

"We can't wait all day," he told me.

"I know, but it's important. It won't be more than five more minutes," I said, hoping I was right.

I held my ground for nearly fifteen minutes until the delivery person walked out and stood by the tarmac with my precious Styrofoam box. In those days, there was no security. I signaled to the pilot, then ran and collected my box.

"We can go now. Thank you," I announced as I climbed the staircase into the plane.

In Brisbane, I deposited the precious blood fractions in a lab at the university. I'd made arrangements for the white cells to be stored there and then sent to South Africa in liquid nitrogen. The only requirements for the red cells were normal refrigeration and wet ice. The South African lab would do detailed genetic analyses on both cell types. This part went smoothly.

Next, I called the airline and told them I needed to return to San Francisco immediately because of my exposure to HBV. My mother sent a telegram, using her physician's credentials, and I was on the next day's flight.

The plasma-derived HBV vaccine had just been approved and was available at UCSF. I received the first two shots of the three-dose series and figured that was enough. I never got hepatitis, but to this day, I test positive for antibody to the hepatitis B core antigen, a marker of exposure to the virus.

Back at Stanford, I received word from South Africa that they were able to do some of the genetic typing but that the leukocytes had been compromised. So the amount of data I acquired was minimal, and I couldn't answer the question of whether there was a genetic basis to susceptibility. It was disappointing, but still an interesting experience.

Cavalli-Sforza was unhappy and complained about my "wasted" experience. I didn't want to go through the whole collection process again. It was a brutal, exhausting, and potentially dangerous undertaking.

"I don't understand why you weren't successful," he said, dismissively. "Bodmer[2] was able to get high-quality samples from the Khoekhoe in South Africa."

Dejected and upset, I debriefed Rose Payne about the outcome and Cavalli-Sforza's response. She was familiar with Walter Bodmer's work and shed some light on Cavalli-Sforza's comment.

She said, "You know, what Luca didn't tell you is that Bodmer had a team of people when he was in Africa. He wasn't on his own, doing everything with no support. You have nothing to be apologetic for, and he should absolutely not be criticizing you for a remarkable effort. You were one person. If it didn't work, it didn't work. Move on."

After that, I decided to limit my time in the lab. I did some additional small projects, like learning DNA-based laboratory techniques and collecting blood samples for mitochondrial DNA analysis from local Bay Area Tongan and Samoan communities. I left Stanford certain that I did not want a career in either science or academia. I had learned by this point never to look back. I avoided returning to Stanford's Medical Center for decades—there was nothing that pulled me there.

Mel and I discussed whether I should return to Illinois. However, his skills were perfect for Silicon Valley and I was finished with academics,

---

2. Walter Bodmer and Cavalli-Sforza were both geneticists with ties to Cambridge University. They collaborated on teaching a genetics course at Stanford in 1962 when Bodmer was a professor there and Cavalli-Sforza was in Pavia, Italy. Years later, they published *The Genetics of Human Populations* (1971), which became a classic in the field. In 1970, Cavalli-Sforza joined the Stanford faculty and Bodmer took a position at Oxford University.

so I resigned. None of my friends and colleagues could believe I'd left a tenure-track position in anthropology. They thought there was only one reasonable explanation: Obviously, I wasn't thinking straight.

Regardless, I finished my postdoc and spent the remaining time exploring my new life with Mel and thinking about what I wanted my future to be.[3]

Lessons Learned

- Be aware of what is in your environment; it just might come in handy when you least expect it.
- Where there is a will, there is a way. If you can define the problem, you can find a solution.
- Always return what you borrow.
- Sometimes it pays to be visibly and annoyingly stubborn, especially if there are no other options.
- Get the key parts of the job done, and then focus on your health.
- Sometimes "failure" happens, but the journey can still be positive.
- Never look back.

3. See "Fame, Fortune, or Family: The Mid-30's Crisis," a contemporaneous account written in 1983 on www.ElizabethReedAden.com (under tab *Mud, Microbes, and Medicine*).

Chapter Twenty-Six:

Career Paths—This Way, That Way, Another Way

George Orwell's *1984* was a landmark book, and for me, the year itself was a watershed. I had to figure out what I would do next. The sexual politics of academia were obnoxious and distracting. Furthermore, I could plot the trajectory of my research for the next ten years, which made it feel programmed and unexciting.

It was time for a career change. What about a business career? Would that offer what I was looking for? At Stanford, I audited courses in the business school and thought maybe it could, but I had no idea which aspect of business to pursue. My brother-in-law suggested marketing.

"Here's what you do, Betsy. Look in the newspaper and find the names of people who have jobs that sound interesting to you. Give them a call."

"I just cold call a complete stranger?"

"Yes. You tell them you saw their name in the newspaper and you'd like to have an informational interview with them. Say you have a doctorate in anthropology and you're making a career change and you'd like the benefit of their experience and advice. Offer to take them out to lunch."

"You mean they'll just say yes?"

"Everybody likes a free lunch. Whatever you do, be sure to set a date, time, and place to meet. Ask if they have a favorite restaurant and, if they don't, suggest a place close to their office. Oh, and most important, during lunch get names of people you can talk to for more or different information. Then use your contact's name when you talk to the person they suggested. That'll get their attention. You're going to be building a network."

"This really works?" I was incredulous.

It really worked.

(Even today, this technique is effective. Instead of calling, like I did in the '80s, I put named references in the subject line of an email or a topic that is right in the recipient's sweet spot. Anything to get their attention.)

At first, I talked with a wide variety of people, including management consultants, marketing directors, and advertising executives. They guided me toward market research, which made sense. There were three options for a market research position: (i) a technical expert who creates questionnaires and conducts data analysis—not enough variety; (ii) a product expert—too much like academia; and (iii) an advertising company position working with a variety of clients and products—bingo! I knew where I wanted to be.

All I had to do was convince an agency that I was a good bet and would need minimal training. Initially, people were intrigued but also confused by my background.

"You're an anthropologist? Like Margaret Mead. Interesting, but why should we hire you?"

"My dissertation identified who gets a disease and why. In market research, you want to know who buys the product and why," I'd answer. "It is the same skill set. I can design a research study, I have data analysis skills, and I've got training in statistics."

It was all true, and I proved it by succeeding in a series of positions I'd never imagined myself taking back when I decided to study anthropology.

After months of networking, I landed a job at Landor Associates, a branding and image consulting firm. The company had designed many now-iconic logos, including FedEx, Levi's, Alitalia, and Marlboro. My position was market research project manager. I worked with freelance market research providers who were recommended by my supervisor. It was a crash course in real-world corporate image, product positioning, and logo design market research tools and techniques to probe what consumers thought about a brand or design. For example, Shell Oil wanted Landor to refresh its gas station design. There were two station types in the US market: a ranch style with a peaked roof and a hexagonal canopy style. The latter was proprietary to Shell, and they hoped to use it as part of their branding imagery. My job was to figure out which direction best matched the company's desired communication goals.

What to do? How to do it? Clueless, I struggled with how to design an informative research study. Asking car owners or Shell customers which design best conveyed the desired imagery was a start, but it wouldn't dissect subtleties between and within the different design elements. Shell was also tweaking other elements, including the logo's color and shape. The problem was complicated, and the research would need to provide clear guidance on these three parameters.

Then I had an insight: Use a series of pictures that became progressively more informative. The questionnaire began with a simple drawing of a building—either the hexagonal canopy or the ranch style. Car owners rated how well each structure conveyed specific attributes, such as trustworthiness, quality, and service, on a scale of one to ten. The next picture had either a brand cue (Shell's existing red and yellow colors or the proposed gray and red) or a category cue (gas pumps). The final picture was a branded Shell gas station, either with a hexagonal canopy or the ranch style.

The research design was novel and gave clear answers that both Landor and Shell Oil executives embraced. Consumers spontaneously associated the ranch style with Shell and said it conveyed service, quality, and trust—desirable imagery. In contrast, open-ended responses to the hexagonal canopy station included "toll booth" and "drive-through Fotomat," communicating the perceptions of fast, self-service, and cheap—undesirable imagery for Shell. Today, you can still see the occasional hexagon canopy self-service gas stations, but the newer stations are all the ranch style, which Shell pioneered.

Based on the success of the Shell research, I did a similar market research project for *People* magazine, which helped them analyze the placement of different elements on their cover. Whenever I see *People* magazine in stores today, I smile because they incorporated and maintained our recommendations.

At Landor, I learned the presentation and communication skills that became critical in my career. In the early 1980s, making a presentation was not simple. Landor employed graphic designers to work on Dicomed machines that cost a quarter of a million dollars.[1] These specialists showed me how to transform a wordy slide into a picture. After we created the

---

1. Approximately to $800,000 in 2026 dollars.

presentation, we sent the file to a company that converted it into thirty-five millimeter slides. It was a cumbersome, time-consuming, and expensive process, but we made beautiful, memorable, and impactful presentations. In time, I became proficient in Landor's style and all was going well.

You might say I had fully integrated into their culture.

Until it became clear that I hadn't.

And I didn't necessarily want to once I realized a person who worked for me earned more than I did. When I confronted my manager, the conversations didn't go the way I expected.

"Well, you don't have formal training in business."

"Really? I have a PhD."

"That doesn't count," she replied.

"How can a PhD not count?"

"Bill has more experience."

"But I'm his manager. Shouldn't I earn more?"

"No. He came in at a higher salary."

Sigh. That's when I learned to ask for a higher salary and more vacation at the very beginning.

Having now done market research for three years, it was also becoming routine, and that meant . . . boring. I don't do boring well. Even though I had pioneered some new market research methods, it was time to move on.

My time in academia had taught me that I was able to go up the learning curve quickly, and working at Landor reinforced that. I could readily grasp about 80 percent of a topic in days or weeks. My experience researching hepatitis B suggested that it took me five years to grasp 95 percent of the material, and another three to five years to reach 98 percent. I'd spent ten years in hepatitis B research, and while there had been some extraordinarily euphoric, exhilarating ah-ha moments, I loved the "mind candy" and challenge of a steep learning curve.

The question was, what would I do next? I was thirty-eight years old when I was offered a vice president position at an advertising agency. I wondered, *Should I take it? Or should I move laterally and find a new learning curve?* The learning curve won. Studying a disease had been interesting, and the newspapers were filled with buzz around the emerging world of biotechnology. It sounded fascinating. I began calling people I read about in the paper and put my networking technique to use. When asked, I was clear about my salary "needs" and vacation expectations.

My next job was at Genelabs as the manager of corporate development, which, to my surprise, included responsibility for developing and participating in licensing agreement negotiations. Who knew? I'd never read a contract, never written a business plan, and never used a spreadsheet. Moreover, much of the science terminology was new, as were the regulatory and clinical requirements for drug development. Perfect.

Central to any pharmaceutical product licensing effort is knowing how a patient could benefit. Where to begin? What is a patient benefit? What about risk? Who could I talk to? I started by going into the labs, where I met and talked with the scientists who described how a molecule worked (or was supposed to work). Then, armed with knowledge about the mechanism of action, I walked down the corridors and into the offices of the clinicians who designed trials to test the drug's benefits. Armed with newfound knowledge, and with help from senior management, we out-licensed some compounds to Big Pharma. Genelabs provided me with a comprehensive introduction to the business, legal, patent, and clinical development aspects of biotechnology, and, unexpectedly, public relations.

In 1990, Genelabs was about to start clinical studies with Compound Q, the company's promising treatment for AIDS—a lethal disease that, at the time, was epidemic mostly in large cities. The regulatory review had stalled, and the company's receptionist was swamped with calls. That's when the CEO asked me to become the company's spokesperson.

With no formal training, I gave interviews to reporters about our promising compound. Compound Q, or trichosanthin, was a protein found in Chinese cucumber roots. The protein destroyed cells with multiple nuclei. This attribute was perfect for treating HIV infection—it was also perfect for inducing abortion, which was the compound's traditional use.

Most often, the articles resulting from my interviews would focus on the protein being a natural product used in traditional Chinese medicine. True. This left readers with the impression that because it had been used for thousands of years, the compound was safe. False. The reporters' subjectivity was similar to what I'd experienced during the Coconut Rebellion in Vanuatu. It was difficult to find an objective and comprehensive description of the compound in the news. At that time, Compound Q had the potential to become an important drug. Genelabs and Sandoz, its marketing partner, supported community-based clinical studies. The drug was never approved.

After three years at Genelabs, a diminished learning curve, and management changes, it was time to move on.

Lessons Learned

- Business is a much cleaner game than academia.
- Do what's right for you but take time to think it through.
- Networking-based sales techniques are effective—get a date and a name, call the name, and use the first person as a reference.
- Ask for a higher salary and more vacation time than you may think is reasonable. You just might get it.
- Recognize what you like and are good at. There is probably a job or career that rewards it.
- If you like learning and exploring, lateral career moves can be more interesting and rewarding than vertical moves.
- Learning makes work fun.
- When something loses its value, it is time to move on.

Chapter Twenty-Seven:

Family Matters

Like my professional life, my personal life never seemed to follow the expected trajectory. Our hope for a family did not go according to plan, and, sadly, I experienced two miscarriages. Mel and I tried a couple of in vitro fertilization cycles, but that wasn't successful either. We also worked with a surrogate for four months until she moved on with her life. That left us with adoption. Private adoption had recently become an accepted and viable option for couples. The process was straightforward. A pregnant woman would go to either a private adoption agency or an adoption lawyer. As part of the interview and screening process, she would explain why the baby was being put up for adoption.

We took this route and wrote a letter of introduction complete with our photograph to market ourselves to a potential birth mother. Our one-page letter was placed in a book with letters from thirty to forty other couples also looking to adopt. The prospective mother would browse these letters until she resonated with a couple. Patty chose us because Mel looked like her son, who had been born when she was seventeen and single.

We flew her to California from her home in Iowa when she was six months pregnant. The introductions went well, and we thought she was funny, smart, pretty, and delightful. She was a court stenographer. Everything was going smoothly, and she even passed muster with my mother. She flew home again, and three months later, we met her at the airport in San Francisco. Patty lived with us for a week until she gave birth, and even helped us decorate the baby's room. Over dinner, we discussed names.

"How about Brandon?" I offered a then-popular boy's name. "Brandon Reed Aden. That sounds great."

"And the initials spell 'bra.' No boy wants to have those initials. Guaranteed," Mel volunteered.

"How about Mikesell?" I offered, remembering how the Ivy League favored students whose first names were also last names.

"My sister's son already has the nickname of Mike."

Not giving up, I suggested Ainsworth, my godmother's family name.

"Seriously? You're going to let his nickname be Anus Worth?" my youngest brother said in disgust.

After all those nonstarter names, someone suggested Marshall. We all liked it. The only downside, which my brother pointed out, was that he might be nicknamed Marshmallow. In exhaustion, we recognized that every name had a downside and decided that Marshall had enough positives to override the possible nickname. Everyone agreed that the baby's name would be Marshall Reed Aden.

Patty's water broke on St. Patty's Day in 1990, and we took her to the hospital where Marshall was born the next morning. She had a natural birth. We brought Marshall home the next day and relocated Patty to a nearby motel.

Talk about learning curves! Mel and I took home a seven-pound mammal with blond hair and blue eyes. He was a small alien, and we were supposed to figure out what he needed every minute of the day. We ticked off a mental checklist. Food? No. Diaper? Sometimes. Cuddle? Often. Movement? Almost always. To survive, Mel and I formed a tag team by saying, "It's your turn."

Although the adoption had gone smoothly up to that point, I struggled with it. It was heartbreaking to take Patty's precious infant. I wrote her a note during Marshall's first night with us. She clearly loved this baby, and I felt terrible. In my letter, I admitted that I was conflicted—on the one hand, he was gorgeous, but on the other hand, I felt like a thief.

Patty was gracious. We found out later that she regretted her decision to put the baby up for adoption. A social worker helped her realize the decision had been best for the child, and she signed the release allowing us to finalize Marshall's adoption.

Since Mel had been the reason she selected us, he was the one who kept in touch. The next time Patty and Marshall met was when she flew to California for his second birthday. Again, it was difficult. This adorable blond, blue-eyed child in the bathtub was calling me Mommy. That was tough for her to hear.

The New, the Unexpected, and the Familiar

Professionally, I was falling into a pattern in which I would get bored every three to four years and push a boundary too far. Usually, there would be a new manager who wasn't willing or able to put up with my outspokenness or honest feedback. I would then be fired or asked to leave. By the time the job reached this point, the feeling was mutual. It would be time for me to find a new challenge in a new environment.

When I was fired by the new Genelabs COO, I was savvy enough to negotiate a severance package. Over my "summer vacation," I explored other opportunities in biotech but wanted another lateral move. I hoped to keep learning and expand my network. My next step was moving into Big Pharma, Syntex Pharmaceuticals, as a new product planning manager.

In 1992, soon after I joined Syntex, Mel and I used our vacation time to return to Malo Island as a family. First, we rendezvoused with my mother and her fourth husband, Robert, in New Zealand. Malaria on Malo was now chloroquine- and Fansidar-resistant, so mefloquine was the recommended antimalarial drug. In Christchurch, before going on to Vanuatu, we all took our prophylactic dose of mefloquine so we would be protected. We flew to Fiji the next day for a two-day visit. That first night, at 3:00 a.m., my mother knocked on our hotel room door. I woke up and groggily staggered to the door.

"Who is it?" I whispered, not wanting to wake up Mel and Marshall.

"It's your mother. Let me in." I opened the door. She was in her pajamas, and she quickly closed the door behind her.

"What's wrong?"

"Robert is having a psychotic breakdown. He said he was going to kill me. When someone says they're going to kill you, you believe it," were her words of psychiatric wisdom.

I moved Marshall into our bed and gave his bed to Mom.

That morning, we discussed what to do. I rebooked our flight to Vanuatu for four days later.

My mother and I took a taxi to Sigatoka and bought some haloperidol to sedate Robert, a strong, skeletal, six-foot-eight, seventy-year-old man. It took five large Fijian men to hold him down while she medicated him. Fortunately, we were at a resort. At the hotel's suggestion, Mom hired Joshua, a rotund, easygoing Fijian. He was described as "the storyteller" and would be Robert's constant companion and caregiver. She hoped Joshua's stories and melodic voice would both entertain and pacify Robert.

With the crisis now under control, she and I flew to Vanuatu. We left Mel and two-and-a-half-year-old Marshall behind so Robert wouldn't feel abandoned.

Upon our arrival in Port Vila, where Ken Hutton now lived, I called my longtime Aussie resource. I told him my mother and I had just arrived and had a two-hour layover before our flight to Santo. Ken had always wanted to meet my "mum," so he drove to the airport and offered to fly us to Santo in his private plane. Mom was afraid to fly in small planes, so we declined. However, he insisted we stay with him when we returned from Santo.

We flew to Santo in a small propeller-driven commercial plane, which Mom didn't like, but it was a lot larger than the eight-seat Islander of my previous trips. After we checked in to the Hotel Santo, I walked down to the Sarakata River, where the small boats were moored, and arranged for someone to take us round trip to Malo the following day.

My mother, the same person who'd sent me to Malo in 1973 with a safari outfit and tons of tetracycline, came prepared. She wore a hat with a mosquito net, a long-sleeved shirt, and pants. There was not a single spot of flesh available for an *Anopheles* mosquito to pierce. As we sat under a banyan tree near the river waiting for the boat to arrive, Mom dictated her thoughts into a microcassette tape recorder: "Betsy and I are waiting for the boat to arrive. It's hot even in the shade. I am looking forward to seeing Malo, where Betsy spent so many years."

The monologue was not unusual, as she kept detailed notes, tapes, and diaries about her thoughts and feelings. She said it was for the book she was going to write—but never did. Thirty minutes of dictation later, the boat arrived and dropped us off at Malahatou.

"We heard you were in Santo and are glad you and your mother were able to come," Vira said as he helped guide the speedboat to the beach.

"I am going to the garden and will make some *laplap* for you and your mother," Voligo said. "We will have a Malo dinner before you return to Santo."

"Thank you. In the meantime, I'll show my mother the Mission House and introduce her to Jole and Vomule."

Several minutes later, Mom and I walked into Avunatari Village. I pointed out the key sights: the graves of Antas and the missionaries' children along with the church, dispensary, school, community well, and football field.

We then followed the road until we reached the single-track spur that led to Jole and Vomule's house. Before walking up the slight hill to their house, I stopped in front of a square patch of broken coral—all that remained of my house.

"This was my house," I said as I walked around the perimeter. "Here's where my bedroom was, and the living room and the spare bedroom." Then I pointed to a rectangle of coral. "There was my kitchen."

"It is bigger than I pictured," Mom said as she stepped into the living room.

Jole came down from his house. He was now a member of the Vanuatu Parliament.

"Jole, this is my mother, Marilyn. I wanted her to see how I lived when I was here."

"We are happy to welcome you." He shook hands with me and Mom. "Please come and have tea with us."

Vomule was her normal warm self with a ready smile and liquid laugh. "How old is your mother?" she asked.

"She is sixty-five years old," I said.

"No. She looks too young and strong," Vomule said.

On reflection, I realized that sixty-five was old on Malo and few people lived that long.

After tea, we walked around. The island looked and felt different. The government had begun subsidizing copra production, so much of what had been bush during my time there was now cleared and planted with coconut palms. Without the dense foliage of indigenous shade trees and vines, the island was much hotter. There was now a radio connection with Santo and Vila. Malo felt updated—less isolated, but also more sterile and less cozy.

We went back to Vira and Voligo's and feasted on *laplap*. At dusk, we returned to Santo.

"All these years, I thought you were suffering when you were on Malo," Mom said as we debriefed over a glass of wine at the hotel.

"I never said that."

"You weren't suffering at all. You were living very comfortably. Everyone is so nice. It isn't what I'd imagined." I was glad that she had abandoned her Freudian view of "savages." Mission accomplished.

We returned to Vila and spent the evening with Ken, who entertained us with stories of the Coconut Revolution and how his plantation in Santo had been returned to its *kastom* owners. He missed Santo, but since his friends were now in Vila, it was still a good life. This would be the last time I saw him.

The next day, I met briefly with Jole's son, Sumbue, who was working in customs. I promised to send him a copy of Bob Rubinstein's dissertation, and also mine. Mom and I flew back to Fiji. Thanks to Robert's breakdown, our planned week in Vanuatu had been reduced to three days.

Back at the Fijian resort, Robert and the storyteller had become an easily recognizable duo. Joshua had Robert under control—just barely. In its wisdom, the hotel's management let Robert have the run of the resort. He ate when he wanted to eat and went where he wanted to go. The haloperidol kept him sedated and docile enough for us to make the flight home to San Francisco. At the beginning of our trip, Robert had signs of cognitive impairment. The single dose of mefloquine exacerbated and accelerated his cognitive decline, a known potential side effect. He never recovered and died eight years later with Alzheimer's.

Me? I went back to work.

Lessons Learned

- Adoption can work, but you have to be aware that emotions might arise that you never expected.
- When someone says that they're going to kill you, believe it.
- A lot can happen in ten years in terms of technology.
- A decade doesn't change your friendships and relationships.
- Modernization is not always an improvement.

Chapter Twenty-Eight:

Just Say Yes to Drugs

Yams are a staple in Vanuatu—plentiful, colorful, diverse, and delicious. It never occurred to me that a company founded on yams would open the door to my career in pharmaceuticals. Syntex chemists, whose roots (pun intended) were in Mexico, discovered that the wild Mexican yam was a cheap and abundant source of hormones, which made birth control pills accessible and affordable. In many ways, the impact of their discovery had fueled the sexual revolution of the sixties and catalyzed the feminist movement of the seventies.

The billion-dollar drug, however, was discovered after Syntex moved the company from Mexico to Palo Alto, with its good climate and proximity to Stanford. Here their expertise in steroid chemistry created the blockbuster anti-inflammatory, analgesic drug Naprosyn.[1] This was fifteen years before I joined the company and one year before the drug became generic. The company knew its sales would fall when that happened, so they were aggressively developing what they hoped would be their next billion-dollar drug.

I took a job in new product planning for the US market and was responsible for the strategic (pre-approval) marketing of three compounds. That sounded good, but I wasn't clear on what that meant. My counterpart in the International Group gave me a crash course.

"We're responsible for the target product profile—the TPP," she said.

"What is the TPP, and why is it important?" I asked.

"Simple. The TPP is the outline for a drug's package insert.[2] It defines the condition or disease treated by the drug, how it's dosed, its efficacy, and

1. Syntex also developed and marketed a faster-acting version, Anaprox (naproxen sodium), that became the over-the-counter analgesic Aleve.

2. The piece of paper that comes in the package with the drug, which most people toss into the trash without reading.

its side effects. The essential patient stuff is on the label. Bottom line, if the information is in the package insert then the FDA says it can be promoted by the marketing and sales teams."

"So, what do we do? What's our role?"

"We guide the team that's creating the new drug with a TPP. Ideally, the new drug will address an unmet medical or patient need. We figure out how to use the available data and estimate the potential sales of the drug," she said.

"And we do this . . . how?"

"We work with the scientists to let them know the dosing regimen that that market will support—the ideal is once-daily oral dosing. You'd be amazed at how hard it is to convince them that people just don't take a drug three or four times a day, or nasal sprays—so we beat on the scientists until they find something marketable. After that, we get involved with the clinicians and clinical trial design."

"Why?"

"We need to know how much better the drug has to be than its competition to be clinically meaningful and generate the sales we expect."

She was speaking the equivalent of a foreign language. I knew drugs went through three phases of clinical studies, but I didn't really understand the differences. Learning strategic marketing was similar to living among the ni-Tamambo. I talked with scientists who were thrilled to discuss their research findings. I asked questions and spent hours with clinicians learning about the clinical trials that were underway. I also read journal articles to learn about the competition—drugs that were already approved and marketed, or drugs that were in clinical development to treat the same disease (indication). It was hard but exhilarating work.

I was assigned to Syntex's most technical and promising new product: a drug for solid organ transplantation. It had an almost unpronounceable name, mycophenolate mofetil ("Myco"). It was an odd drug for Syntex and was the brainchild of Anthony Allison, a celebrated medical research scientist.[3] His insight was to make a prodrug of an old compound that had too many side

3. Ironically, I'd met Allison while I was at Genelabs; the company had interviewed him to be the VP of research. And from my anthropology days, I knew of Allison's discovery that people with one copy of the sickle cell gene were protected from cerebral malaria while people with two copies of the gene usually died in infancy because their red blood cells could not carry sufficient oxygen. This was the first clear example of a "balanced polymorphism" in which the heterozygote (one copy of a gene) has a clear survival advantage over either homozygote (two copies of the normal hemoglobin gene or two copies of the sickle cell gene). Allison was a pioneer in the field of medical genetics. During my one-on-one interview with him, I asked if he knew Barry Blumberg. He said he knew him well and had suggested that he use the blood from a multiply transfused patient to discover new antigens. This was exactly the approach Blumberg used to discover the Australia antigen.

effects to be clinically acceptable. A prodrug becomes pharmacologically active only after it has been metabolized (acted on) by specific enzymes in the intestines or liver. Allison reasoned if the patient's intestine activated the drug, the therapeutic dose would be lower and the side effects would be reduced.

When I arrived, Myco was in early clinical trials. It had been tested in patients with new kidney transplants and in patients with untreatable acute organ rejection. Both studies had produced positive results. I joked that I had been assigned to the drug for two reasons: first, because I had a PhD, and second, because I could spell "T-cell" and "B-cell," key components of the immune system. At the time, I knew little about clinical trials or drug development and even less about immunology, solid organ transplantation, or pharmaceutical marketing.

Back on the learning curve, I used my Malo Island skill set and went to professional society meetings. I hired Joe Twist, Vira's equivalent, to introduce me to key people in the transplantation community.[4] He helped me formulate questions, and he explained how the system worked from organ procurement to transplant to patient management. We talked with many different surgeons and nephrologists, some of whom were loosely affiliated with the company. Syntex was new to this field, and the marketing representatives from the other companies—like Fujisawa (now part of Astellas) and Sandoz (now part of Novartis)—were all men. By contrast, the Syntex team was mostly women, and we were developing a promising new drug. The transplant surgeons and physicians wanted to talk with us. For example, Thomas Starzl, the surgeon who pioneered liver transplantation, approached Syntex about using Myco in an experimental procedure that would transplant a pig liver into a patient with liver failure and AIDS.[5] However, Syntex rejected his request. We relied on the advice and expertise of Hans Sollinger, a pioneer of simultaneous kidney and pancreas transplantation and a friend of Anthony Allison's, to guide us through the preclinical studies and clinical trials.

It turned out I didn't have to be a genius to succeed with Myco, but I *did* have to be a good listener. Everyone had an opinion about how to use and promote the drug. My job was to sift through their ideas and determine

4. Joe Twist had been the strategic marketing director for the drug Sandimmune (cyclosporine), which became the mainstay of solid organ transplantation and made possible cadaveric transplants. The drug revolutionized solid organ transplantation and became the best-selling product in Sandoz's portfolio.

5. A photograph of me (as an "unidentified person") with Starzl appeared in the American Society of Transplant Surgeons' (ASTS) magazine *Chimera* soon after I attended my first ASTS meeting.

which made sense. After three years of working with leading transplant surgeons and physicians, I knew their scientific and clinical interests. I figured out who had good marketing sense and who gave me solid technical and/or clinical advice. It was like playing solitaire with the cards face up. All I had to know was what I wanted to accomplish.

Risks and Rewards

There were some critical inflection points during my three years of working on Myco. The drug's development timeline was years behind schedule because the company couldn't find a reliable manufacturer. Under managerial and financial pressure, Syntex took calculated risks to accelerate the pivotal Phase 3 trials necessary for FDA market approval. Everyone understood that the company's fate depended on Myco.

Risk #1—Trust in Small Numbers: The first clinical trial (Phase 1) had seventeen patients, who were treated with either two or three grams of Myco daily starting when they received their new kidney. This study showed that Myco reduced the number of acute rejections by 40 percent. Great, but there was not enough information from this trial to determine the effective dose.

Risk #2—Skip Phase 2: Phase 1 is usually followed by a Phase 2 trial that defines the effective dose, since Phase 1 is more about safety and tolerability. Syntex was desperate, so the company skipped the Phase 2 trial and rushed into multiple Phase 3 trials.[6] But the deck was stacked against us.

Risk #3—A Phase 3 Protocol Using an Unapproved Drug: Transplant patients were being treated with an antibody concoction (Minnesota antilymphocyte globulin, or MALG) developed at the University of Minnesota to temper the patient's immune response against trying to reject the transplanted new organ. MALG had never been approved by the FDA but, at the time, was still part of the standard protocol. Then, three things happened: (i) The first monoclonal antibody, OKT3 (muromab-CD3), was approved; (ii) OKT3 could be used to replace MALG; and (iii) The FDA halted use of MALG while pursuing criminal charges against its developer.[7] Three months

6. At the time, Naprosyn had come off patent and the Syntex's sales were falling dramatically. Syntex lost more than 50 percent of Naprosyn (naproxen) revenue within the first few months after the loss of patent protection in December 1993 and immediate launch of generic competition. At the time, Naprosyn represented nearly half of the company's sales.

7. L. G. Wilson, "The Crime of Saving Lives: The FDA, John Najarian, and Minnesota ALG," *Archives of Surgery* 130, no. 10 (1995).

into Syntex's pivotal registration trials, with twenty-seven patients treated, Syntex had to replace MALG with OKT3, losing precious time.

While the clinical and regulatory team sorted out the MALG hiccup, I was tasked with creating a Scientific Advisory Board (SAB). I wanted a diverse group of physicians and surgeons, at least one of each for the major transplanted organs—a Noah's Ark of transplant clinicians. International representation was also important. Hans suggested names. Ellen, the vice president of clinical development, and I flew around the country interviewing prospective candidates. Besides the obvious criteria of being smart, articulate, and knowledgeable, a key consideration was whether we'd want to have dinner with them. Surprisingly, many candidates failed because they lectured rather than conversed.

Our first SAB meeting was in the summer and began with an update on the just-completed Phase 3 trials. There were smiles all around. Myco was the kind of drug every pharmaceutical person dreams about. It unambiguously passed every test in preclinical studies. Each clinical trial generated clinically meaningful and statistically significant results. The drug just worked. No special subset analyses were needed. No excuses. Syntex would be saved. But then one more risk arose.

Risk #4—Assuming Efficacy in Small Studies Is Reproducible in Large Trials: Bernadette, the Syntex clinician who ran the Phase 1 trial, informed the group that the number of acute rejections in the Phase 3 studies was higher than expected. No one knew why, and no one knew which treatment group(s) had more rejections. The clinical trials had assumed a 40 percent reduction in acute rejection, the rate that had been observed in the Phase 1 studies.[8] Assuming the same 40 percent efficacy, we needed 125 patients per group to reach statistical significance.[9] If we stopped the trial as expected, it was a virtual certainty the drug would fail. Her news was bad . . . disastrous. What to do?

"What would be a clinically meaningful reduction in acute rejection?" Bernadette asked the SAB.

"Twenty percent," one of the clinicians answered.

---

8. This was the first time that the FDA agreed to use reduction in acute rejection as a surrogate endpoint for survival benefit. This meant that positive results could be seen within the first three months post-transplant rather than enrolling lots of patients and seeing a survival benefit after one to two years. It was a huge step forward in bringing new immunosuppressant drugs to the market.

9. This assumed there would be fifty fewer cases of acute rejection in the Myco group versus the patients receiving the then-standard-of-care treatment.

The biostatisticians quickly calculated the additional number of patients needed in each arm of the study to reach this threshold.

"How long would it take to recruit these additional patients?" Bernadette asked.

"At least another three months," the clinician replied.

We were losing more time, and the company's fate hung in the balance. Would senior management agree to extend the trial?

Interlude—Taking Notes

I learned from this experience to volunteer for note-taking—it meant that I learned the material better than anyone else and controlled how the notes were written. I promptly became the go-to person for Myco-related new product marketing tasks. There was no meeting I skipped, no conversation I avoided, no training course I declined. I took every opportunity that appeared and asked for more.

During the clinical trials, I worked with transplant surgeons who were at the top of their field, internationally renowned, and among the most demanding people in the world. I was even given the opportunity to be in the operating room while one of them performed a heart transplant—an honor and a privilege.

I had no idea what to expect when a nurse came up to the observation area and asked if I wanted to have a closer viewpoint during the procedure.

"Sure," I said. "That would be great."

I changed my clothes, put on a gown, scrubbed up, and was admitted to the operating floor. I had a clear view of the patient's open chest and watched the surgeon as he prepared to remove the old heart. I was less than two feet away.

This patient had already undergone one heart transplant and was now getting a second one. As the surgeon cut the aorta, it looked as though he was cutting along the "dotted line" where the heart had been sutured.

When the old heart came out, I held my breath. The man's chest seemed empty. Next, a nurse opened a container and took out a gray lump of meat. It was the new heart. I thought, *There is no way this is going to work.* The meat in the supermarket looked fresher.

Undeterred, the surgeon sewed in the replacement heart. And nothing happened. I uttered something like, "This'll never work," but was reassured that it would all be fine. Then the miracle happened. Blood flow to the heart was restored. The surgeon massaged the organ, and then . . . it beat . . . on its own.

Those last minutes were terrifying and then joyous. The next day, I visited the patient in his room. He was on his way to recovery.

What's in a Name?

As the US product manager, I found the Myco experience fun and exciting. I had carte blanche to do what I thought would support the drug's premarketing. It was also my responsibility to find a brand name. We'd looked at hundreds of names, most of which were awful. Finally, I hired a local San Francisco advertising firm that suggested Cellect and Axcept as possibilities. My counterpart on the international product side offered an embellishment.

"While I was skiing on my NordicTrack this morning, I thought we could combine the two best names. How does CellCept sound?" my colleague asked.

"Sounds great."

We had a name everyone liked and one we could trademark. We next agreed on the drug's trade dress—the color and shape of the tablets.

Our weeks were filled with creating symposia on key topics, attending professional meetings in exciting venues, taking physicians out to dinner, going on rounds to learn about patient needs, talking with Medicare officials about reimbursement, and meeting with organ procurement organizations. It was important to understand each of our different users, prescribers, and payers. Syntex provided the budget, and management gave me a free hand to invest my time in those early premarketing efforts.

I was traveling 60 percent of the time, which took some juggling, as Marshall was about four years old. Mel was Marshall's primary caregiver—which had an effect, too.

"Where does Mommy live?" Marshall asked at one point.

"She lives with us," Mel replied.

"No," he said. "No, she doesn't."

I wasn't around very much, but fortunately, Mel helped make it all work.

Call It Quits or Increase the Power?

It was a wonderful, challenging, and exciting time up until that first SAB meeting, when we started to wonder—was Bernadette's data Myco's death knell?

The US senior management team wanted to cut back my budget while corporate senior management decided whether to proceed with additional enrollment in the Phase 3 trial. I convinced the US executives not to change course based on the strength of the available clinical and preclinical data. I was confident the drug worked and was highly invested in every step of its development.

Within days, the decision was made to continue the trials, which were completed for the second time in spring 1994. By now, however, the company's revenues had declined. Generics had taken nearly 90 percent of branded Naprosyn's market share. Syntex was for sale—that was the gossip. The company's cash cow had been slaughtered.

Still, we assumed that at least two of the three pivotal trials would be positive, so we held our first product launch meeting—at Chicago's O'Hare Airport, which was centrally located for some of our key transplant clinicians.[10]

During lunch that day, May 4, 1994, I picked up a voicemail. The message was from my immediate supervisor: "Roche has submitted a tender offer to buy Syntex. The offer is contingent on the CellCept clinical trials being positive." In shock, I hung up the phone. I had no sense of what this meant for me, the drug, or the meeting. I did the only thing I could do . . . I went back into the room and continued the meeting. I said nothing until after the meeting, when I confided in the other Syntex employee who was present.

The following weekend, the clinical and biostatistics teams worked non-stop. On Monday morning, the CellCept team learned that all of the clinical trials were positive. I was told the results four days before the annual transplant physicians' and surgeons' meeting. Using money I had on my corporate card, I sent a gift-wrapped bottle of champagne to Hans—a thank you for successfully guiding the program.

The gift, however, arrived with the cryptic message "You're about to be a father again."

That night, he called me from his car. In his German accent, he said, "Mary and I are on our way to the mountains. I don't think that I am about to be a father again."

"Did you open the gift?" I asked.

---

10. Since there were two doses of Myco in the Phase 3 study, the FDA demanded that instead of statistical significance being at $P \leq .05$, it be $P \leq .025$ (a Bonferroni correction). The FDA set a very high bar for efficacy.

"No. I gave it to my office mate."

"We can talk at the meeting," I offered.

"Okay," he replied.

I found Hans and Mary at the meeting soon after Syntex announced the CellCept Phase 3 clinical trial results.

"It was a bottle of Dom Pérignon," I told them. "Hans, you are the father of a new drug!"

"I'm getting my present back from my office mate," he said.

Sweet success!

Lessons Learned

- Listen—the answers are out there. Different people have different strengths, so learn who is good at what and follow their suggestions.
- Drugs that are going to be approved just work.
- Commit 100 percent; do everything.
- Always volunteer and always take notes. You'll be invited to the meetings, become an expert in the material, control the information, and move up in the organization quickly.
- Take every opportunity the company offers: travel, meetings, courses. Build your skill set.

Chapter Twenty-Nine:

It's Basel, Not Basil, Not Basal

When Roche took over Syntex, my job and supervisor changed. My new boss insisted that I be in the office until 5:00 p.m. every day. This was a problem for me.

From 1992 through 1995, Marshall stayed at home with our housekeeper two days a week and spent the other days in daycare or preschool. If I was in town, I would leave the office around 3:30 p.m. to pick him up just before the facility closed. I had to drive forty-five miles in rush hour traffic from Palo Alto to his preschool in Berkeley. My new boss said that was unacceptable.

"Is there something in my work that's a problem?" I asked. "Have I missed any deadlines? Is there a problem with the quality of my work?"

"No," he admitted. "But you should be here."

That was ridiculous. "Why? Why should I be here?"

He had no reply. My work never suffered, and I accomplished more than most people in the company did, but there was that old notion that everyone should stay until five o'clock even if they were doing nothing, even if it was just for show. I thought, *This is not about what works for me. It is about what works for the optics.*

It didn't work for me, so I ignored him and kept my schedule. My hours were generally long, and I could easily continue to work from home. A few weeks later, he was no longer my manager.

CellCept was transitioned to Roche's New Jersey site by the end of 1996, and with the launch plan we'd developed—pioneering use of the nascent World Wide Web, plus our marketing director's expertise—it became a multibillion-dollar drug. The SAB we'd put in place advised the company for the next decade.

As CellCept transitioned, I was assigned to learn about genetics, personalized medicine, and gene therapy. Since I'd demonstrated I could spell "T-cell" and "B-cell," it was assumed I could also spell "DNA." Understanding the genetics of disease susceptibility and gene therapy was a new area for the pharmaceutical companies. Fortunately, I could already spell "gene."

While others worried about their future role or next job, I was optimistic. I had survived multiple rounds of "reduction in force" and thought it was best to keep my head down and learn. Rather than worry about what was next, I transitioned from Syntex's Mac world to PC Roche. At the office, I became an expert at using MS Office programs. I explored PowerPoint's capabilities to convey what I had learned from the scientific meetings I'd attended on gene therapy and genetics.

Despite the sense of being on a sinking ship, I enjoyed learning and exploring a new field. The Human Genome Project was laying the foundation for personalized medicine—prescribing medicine(s) that works for an individual based on their genetics and/or physiology. Notably, there is a Goldilocks effect for most prescription medicines: They work well for about 40 percent of people and contain either too little drug or too much drug for the remaining 60 percent. I understood the potential therapeutic impact of these new approaches to treatment—even in 1996, it was obvious.

Still, I was nervous when, one afternoon, I was summoned to meet with Roche Palo Alto's new site head. I fully expected to be fired. Instead, I was thoroughly charmed by Ken, a handsome, silver-haired New Zealander. He asked about my experience and was impressed that I had worked in biotech, which was unique at Syntex. He said that Palo Alto would become Roche's biotech campus—innovative and lean. He asked me to write down my thoughts about what the new orientation could look like and send them to him.

I spent that weekend writing and rewriting my ideas for this new Roche site. I knew my future depended on what I wrote in my note and how I wrote it. I had to find a way to be honest, tactful, and succinct. It wasn't easy. My one-page note went to Ken that Monday morning. Then, I waited.

When Ken held a company-wide meeting, he announced that Syntex would be divided into two business units. A few weeks later, I was summoned for another meeting. This time, he offered me a promotion. I became head of strategic marketing and business development for the inflammatory diseases. There were rules for headcount and budgets. Initially, I had a headcount of eight and would get to choose my team.

At that time, I knew clinical development well but had only a vague sense of how medicinal chemistry, drug discovery, and preclinical development functioned. One afternoon, I met the heads of chemistry, biology, and preclinical toxicology for our business unit. "Would you mind giving up four headcount so we could do high throughput screening?" the chemist asked.

"What is high throughput screening and what's the benefit?" I asked.

"It is how we screen chemical compounds to find leads for new drugs. Remember, we are getting extra bonuses if we find important new drugs."

"Okay." I gave him four of my employee slots without a fight. It wasn't a hardship, since I thought my group could function effectively with fewer people. And, it seemed important—it was a means to find new drugs to benefit patients.

Developing attractive new drugs was important to employees because of additional ways to earn bonuses intended to incentivize productivity among the remaining Syntex scientists and clinicians. Everyone in the business unit received an additional reward equivalent to their maximum annual bonus each time a drug was accepted into Roche's Development Portfolio. My boss, the head of the Inflammatory Diseases Business Unit, asked me and Hal, a biologist, to develop a way to prioritize our preclinical compounds to maximize our extra bonuses.

I searched the literature and talked with consultants. There was nothing on portfolio analysis for preclinical products. For the next three months, my group and I figured out how to develop an objective approach to replace the gut feeling that was the usual basis for prioritization. We met with chemists and learned how they evaluated risk, what classes of compounds had a track record of success, and so on. We then distilled their risk criteria into a set of questions with multiple-choice responses.

When we thought we'd captured the main risk and success criteria, our little Strategic Marketing Group tested the instrument with each department (clinical, biology, chemistry, manufacturing, and sales forecast from our group) and made refinements where warranted. For the commercial assessment, we used key elements involved in creating a product's forecast: size of patient population, unmet medical need, competition, etc. This was the portfolio prioritization project.

We had the tools to prioritize, but did we have a uniform scheme to classify preclinical projects? The simple answer was "no." Fortunately, there was a new head of global research who wanted more transparency in research and formed a diverse team, led by an experienced chemist. That team collaborated with my team to develop a classification scheme for all preclinical products.

Now, for the first time, we had the tools to categorize projects by stage of preclinical development and prioritize them within the stage—in other words, tools to manage Roche's research portfolio. Roche had projects in ten different therapy areas spread over six sites[1] with more than 125 compounds in preclinical development. I flew around the world twice in two months, first to collect the project and prioritization data and then to report the results. I worked with the biostatisticians on ways to analyze the data.

The next step involved working with the six research site heads to integrate the portfolio results into their annual goals. In less than a year, Roche Research had a robust management tool that allowed the head of global research to effectively, efficiently, and transparently manage the organization.

I learned that asking people questions was one way to get them to focus. I'd say, "This is where your portfolio is today. Where will it be next year?" Or, "How many compounds are you going to move forward into the next stage? Which ones, and why?"

For the first time, compounds were being compared across sites and therapy areas. Never before had objective measures been used to compare projects. Compounds had high preclinical and early clinical failure rates, and we hoped that prioritization could improve our success rate.

One wrinkle: Our approach was threatening to the research site heads, who preferred to manage their portfolio by gut feelings.

"Has that worked?" I asked them.

The answer was . . . it was hard to tell. In other words, Roche's multi-hundred-million-dollar research organization was run by instinct. To the research site heads' chagrin, their gut-based assessments only semi-aligned with the new analytics. It was an inefficient and opaque process. It was time to change—and I said so.

My words were not well received. The site heads realized that if projects could be successfully compared and rank-ordered based on value, market size, and likelihood of success, it would increase transparency and performance accountability. They wanted nothing to do with portfolio management.

I pressed on. We *would* do it.

---

1. Palo Alto, California; Nutley, New Jersey; Kamakura, Japan; Basel, Switzerland; Penzberg, Germany; and Welwyn, UK.

They protested. It was a stupid thing to do, they claimed. It was useless. Meaningless. Except it wasn't—and when they saw the results, even the skeptical among them had to admit that it worked. Yet, resistance remained so strong that there were times during the data collection phase when site heads tried to manipulate the answers to the multiple-choice options as we assessed individual projects.

"No, no, no," they'd say. "This drug gets a different score, much higher than you are trying to give it."

"You're telling me a large complicated compound is really easy to synthesize?" I would reply, looking at the structure of the molecule. "How many steps are involved?"

"Fifteen."

I'd look at the question and the options, which clearly labeled fifteen steps as expensive and of moderate difficulty. "That is not an easy process," I'd push back. Then I'd ask, "Is a molecule with five steps easier to synthesize?"

"Yes, but none of our molecules has five steps. Most have more than twenty steps. By comparison, this molecule *is* easy."

"So within your portfolio, it may have a higher priority," I would counter. "But we also have to look at all of the company's projects and consider where to allocate the resources. We need to consider the market size, unmet medical needs, and potential health care savings the compound might have. If everything is equivalent, then a five-step product would be cheaper to produce and could be more profitable."

This was not what the research site head wanted to hear, but it was true.

Before I went to a site, I educated myself about the therapeutic area and studied the profiles of the different compounds and the diseases they were designed to treat. It was important to be informed and able to counter arguments, if needed. It was fascinating to learn about new diseases (maybe I *should* have been a physician), medicinal chemistry, and the limitations of animal models. I became familiar with a range of oncology drugs, orally active antifungal agents, a possible long-acting treatment for hepatitis C infection, and many compounds that never made it to market.

Visiting the different research sites was its own form of culture shock, like discovering that the fruit we live on in California was not part of a Welwyn breakfast or that walking through the lobby of a Tokyo hotel in a bathrobe to get to the swimming pool was beyond gauche. I enjoyed traveling and interacting with people from different cultures.

At the global prioritization and goal-setting meeting, I provided the research site heads with slides and walked them through the data. I said, "These are the parameters you should consider when you define your objectives for the next year, this is how you will be measured, and this is your portfolio. How you prioritize is your decision, but these are the projects where the chemistry and biology have the greatest chance of success. These are the ones with the highest commercial potential, and these have clear established clinical endpoints."

Soon, they realized they could use this information to be more successful and stopped fighting. The decision to fund a project was still the research director's, but it could now be made on a more transparent and informed basis. It was the beginning of a new era of research management.

And a new era in my family's life. Based on my portfolio and research management work, I was asked if I'd be interested in moving to Basel and becoming head of global research strategy for Roche.

Mel said that the best part of the Roche acquisition was they were based in Europe. He had grown up in Italy. So, when I was asked whether or not I would move to Basel, I said yes on the spot.

Lessons Learned

- Don't be afraid to ignore work requests that have a significant effect on family life just for the sake of "optics."
- Take on projects that haven't been done before and get input from the experts.
- Before you can improve, you have to measure what it is that you want to improve.
- Test and refine questionnaires.
- Transparency and objectivity are rarely welcomed as replacements for gut feelings.
- Management will accept new tools once they're proven to work.
- Changes in management offer multiple opportunities—take them.

Chapter Thirty:

From Berkeley to Basel

I moved to Basel in early 1999, six months ahead of my family. Mel stayed with Marshall until he finished third grade. Roche Basel was a different world. I didn't know what to expect as the train traversed valleys and went through tunnels from one side of an alp to the other. The snow-capped, glacier-carved mountains were breathtaking.

Every year, Roche held a three-day managers' meeting with their top 250 executives. The company then had 55,000 employees. I was among the elite attending the meeting at Interlaken.

I was assigned a room at the Victoria-Jungfrau, where much of the meeting would take place. At the first *apero*, pre-dinner drink, I wore my business travel suit—a St. John Knits skirt, blazer, and top that always emerged from a tightly packed suitcase wrinkle-free and ready to wear. I entered the ballroom and surveyed its contents. A swarm of men and a babble of voices, most of them not speaking English. I saw no familiar faces and no other women. Later, I learned that two other women had been present. Once in the ballroom, I did what anyone would do—I headed for the bar.

There are advantages to being a woman in that situation. Men went out of their way to introduce themselves, talk with me—in English—and give me tips on how to navigate the event. Many of them had a marvelous sense of humor and enjoyed a party. By the end of the first evening, I had acquired a small group of acquaintances.

Each attendee's dinner venue on the second night was determined by raffle. We all drew tickets specifying that we'd eat in one of three dining rooms. My ticket placed me with the CEO of pharma and the chair of

Hoffmann-La Roche.[1] I entered the room. There were three tables. The CEO was at one and the chair was at another. I headed for the third table, planning to make new friends and have a good time. From my vantage point at the third table, I watched as numerous executives kowtowed to Chair Gerber.

"Who would you like to have sit next to you, Herr Gerber?" one general manager asked.

This went on for a few minutes. Meanwhile, I settled into my chair with my back to the prestigious tables. Then there was a tap on my shoulder. "Herr Gerber has invited you to sit with him," one of the Swiss managers said.

Herr Gerber had spoken and, as the only woman in the room, I gave up my "fun" seat in the safe harbor and was escorted to the vacant seat on Herr Gerber's right. But dinner with Fritz Gerber was not the nerve-racking, treading-carefully-on-pins-and-needles event I had anticipated. Instead, he was charming, delightful, and open without pretense. I told him about my involvement with CellCept, which led to a conversation about the Syntex acquisition—a phenomenally good deal for Roche because they broke even on their $5 billion investment within twelve to eighteen months.

"The negotiations were complicated, but we had a secret weapon," he said.

"Which was?"

"Paul Sacher. He is a composer and the husband of one of the Hoffmann heirs. Did you know that Organon was also bidding aggressively for Syntex?"

"No, I didn't."

"They also saw the potential of your drug. But Herr Sacher participated in the final negotiations. As a composer, he has a very fine ear for nuance. During a break, he said, 'They will take $24 a share. I hear it in their voice.' We made the offer, and that was the final price."

After the meal, a waiter came by with the cigar cart. My brother-in-law had educated me about cigars so I would know what to buy for him in the duty-free stores. Herr Gerber and I got into a discussion about Cuban cigars and the pros and cons of Cohiba versus Montecristo. I wondered, *Do I dare smoke a cigar?* It was an outrageous thing to do. Mentally, I hemmed and hawed. When the truck came to him, he chose his Cohiba. It was now or never. I chose now.

"Would you mind if I had a cigar as well?" I asked Gerber.

"Of course not," he said.

---

1. At the time, the formal company name was F. Hoffmann-LaRoche AG. Today, it operates as Roche Holdings AG. The company was founded by Fritz Hoffmann-LaRoche in 1896.

I chose a Montecristo No. 2 and savored the phallic symbol, puffing away as Blumberg had taught me years earlier. Soon, everyone in the room looked on in astonishment and disbelief. There were advantages to being a Californian in Switzerland. Nobody knew what to expect. I was different, and that translated into fewer constraints and greater freedom.

Saying Yes (and No)

My career at Syntex and Roche was unexpectedly meteoric. It wasn't anything I sought; it just happened. I worked hard and did quality work, but I wasn't ambitious. My mother would joke that "[I] didn't sleep my way to the top. [I] got there by not sleeping." There was a lot of truth in that statement.

Within seven years, I'd risen from being a new product planning manager to a global senior vice president in one of the world's largest and most successful pharmaceutical companies. I'd been the only person to sit on all the key product decision-making committees and make recommendations for licensing and acquisitions. I was respected for my analytic skills, honesty, humor, work ethic, and directness. It had all happened because whenever a door opened, I walked in and said yes.

When the new CEO of pharma took over, however, I heard a "no" coming my way. I'd had many previous interactions with him, and they had not been good. We'd disagreed about the value of two competing products: one developed by the scientists at Roche Basel and the other by the Roche Palo Alto site.[2] He thought I favored the Palo Alto one because I wanted a bonus. I didn't. I thought it was the better product. He was also angry that I'd fired someone he liked a lot but who I thought was incompetent. He respected what I'd done, but he also made it clear that if I stayed, my role would be diminished. I accepted that and knew it was time to leave the company. My decision was made easier because Mel had been commuting to California every two weeks. He wasn't thriving, but Marshall was.

When I left, I was the highest ranked woman in the Roche organization. I continued working hard up until my last day. When I left Roche, the boss who fired me said he appreciated my dedication to work and my sense of humor. I would never undermine a project, and people found that odd. I didn't play games or waste my time. Some of that came from my time in

2. Neither product made it past Phase 2 because of unexpected side effects.

New Hebrides. I'd also grown up dealing with a complicated, contradictory mother who would pull me to her and then push me away. Mixed with her four husbands and my three baby brothers, that concoction could have thrown me. Instead, I chose to navigate it. Crashing was not an option.

And there was Mel. He never pushed or pulled me in multiple directions. He always supported what I wanted to do. Fortunately, his employers let him work flexibly and unconventionally. He regularly (albeit infrequently) commuted long distances to work by plane. The information age made it possible for him to be a productive hardware designer while following and supporting me everywhere.

I look back on those days with pride, both in what I achieved and in the fact that Mel and I experienced so much on my rollercoaster career ride and our life in Europe. It was never explicit, but I knew I had to constantly work harder, do more, be better, and prove myself worthy of my position. I did it because it was rewarding and exciting to make something new.

The first portfolio process that encompassed Roche's six research sites was implemented in 1998, and more than a decade later, I was told Roche was still using it. I'm proud of that achievement. I've also seen some of the slides I created used in corporate presentations.

My time at Landor taught me not to rely on written words but to complement spoken words with pictures. I also learned how to put science and technical concepts into diagrams. This skill was useful and important in a multilingual, multicultural environment. An unexpected benefit was that it forced me to clarify my thinking. If I couldn't draw the relationships, then I didn't fully understand them. One of my colleagues took my work home and his teenage daughter looked over his shoulder at the presentation.

"Dad, is this what you do?" she asked.

"Yes," he replied.

"For the first time, I actually understand what you do for a living. It's interesting."

He was stunned and began to realize how much more effective and unambiguous pictures could be than his copious word-based slides.

When I moved on from Roche, I figured I'd see what life had to offer now, since there was no looking back.

Lessons Learned

- Cigar smoking is everything it's cracked up to be and can be a powerful tool in the right situation. Cigars give you power.
- The only person you can control is you.
- The first-mover advantage is real. When there are no rules, you get to write them in your favor.
- If you can't draw or diagram the concept and have to rely on words, you don't completely understand it.

Chapter Thirty-One:

Beyond Drugs and into New Medical Frontiers

Human embryonic stem cell (HESC) research was the next headline in my life. It was a new area and got negative public attention in 2001. Headlines screamed about scientists playing God, and many sectors of the public believed that to be the case. Geron, the leader in HESCs, was a forty-minute commute from our Berkeley home. As the senior vice president of regenerative medicine, I was responsible for the program. I knew nothing about stem cells but thought it could be an important therapeutic paradigm. Another challenge and a completely new area—more learning curves!

I educated myself by talking to both academics and people inside the company. My major accomplishment during the nine months I was there was to renegotiate the technology license agreement with the Wisconsin Alumni Research Foundation (WARF).[1] Geron had licensed the rights to six cell types derived from HESCs. Geron presented WARF with development plans that were substantially identical for four of the six cell types. WARF then exercised its contractual right to audit the company. WARF's director and his counsel came to Geron and met with the CEO and lead scientist. Annoyingly, I was not included. The meeting did not go well, so the WARF team left

1. Inventions by University of Wisconsin scientists are licensed to WARF, whose goal is to commercialize these inventions. James "Jamie" Thomson was the first to isolate, derive, and grow HESCs. The blood thinner coumarin was discovered by a University of Wisconsin biochemist, who licensed the invention to WARF and called the compound Warfarin, which became one of the most profitable inventions for WARF.

abruptly—and early. The following day, I learned WARF had filed a lawsuit against Geron, and Geron was planning to cross-file at the end of the week.

"I'll go to Madison. You have nothing to lose," I said.

"I guess that's true. Fine. See what you can do," the CEO said.

I caught the next flight directly from work. I called Hans, a professor of surgery at the University of Wisconsin, Madison. We'd remained friends after my transplant days.

"Hans, I am arriving in Madison this afternoon. What do you know about WARF?"

"They have a new director. He is a good guy. What's up?"

"I need to sort out a way to work together with WARF on human embryonic stem cells. We had a meeting with them and it was a disaster, so I'm arriving at 4:30 p.m. without even a toothbrush. Is there someplace I can stay? Any suggestions?"

"I'll book you a place. I can pick you up at the airport and take you there. The place is near the university."

I met with WARF's director later that day, and he was reasonable. We hadn't completed our discussion, so he invited me to dinner at his home. That evening, we found a solution we both could live with: Geron would retain exclusive rights to three HESC cell types (neural, cardiomyocytes, and islets) and nonexclusive rights to the other three. Geron accepted the conditions, the Board approved, and the lawsuit was avoided.

The conditions and diseases that could potentially be treated by the cell types Geron retained included spinal cord injury, Parkinson's disease, ischemic heart disease, and type 1 diabetes (via islet cell replacement). The company was not actively working on the islet cells, so I took steps to create a consortium to "cure" type 1 diabetes with HESC-derived islet cells; this consortium included Geron, nonprofit organizations, and leading academic researchers.

One cool part of my responsibilities was overseeing research at the Roslin Institute. Geron licensed its cloning technology patents. On a business trip to Edinburgh, I met Dolly, a sheep who was the first mammal cloned from an adult somatic cell, and her creator, Ian Wilmut.

Nine months into the job, it was over. Geron needed to cut its expenses. I was expendable and was fired the day before I was to spend three weeks in Edinburgh working with the Roslin team. Mel had made arrangements to work at his employer's office at Leith Links, site of the oldest golf course in

Edinburgh, and Marshall would be in summer camp in Austria.[2] As part of my severance, Geron paid for all my expenses in Scotland. It was a wonderful break, and I spent my time indulging my inner Dan Brown at Rosslyn Chapel and horseback riding through the Pentland Hills.

Out There on My Own

When we returned to the US, I had time to reflect. Roche had shelved a promising drug to treat Alzheimer's disease. I wanted to resurrect the drug and realized I could do so without a license from Roche. I created Renaissance Medicine and enlisted colleagues to help with the business plan, clinical plans, manufacturing, chemistry, and logo. I met with venture capital firms, but they backed away when they realized there were no patents and we would have to rely on statutory market exclusivity. Ultimately, Renaissance Medicine was acquired by Alfa Wassermann Inc. (AWI) for enough money to reimburse my friends for their time and expenses. As part of the deal, they agreed to help move the Alzheimer's drug forward[3] if I became AWI's chief operating officer. There was only one answer: of course.

Alfa Wassermann built and sold two products: a benchtop clinical diagnostic instrument and sophisticated separations equipment. It was located in New Jersey—and there was another learning curve. I had never managed a company. I knew that the FDA was focused on quality but didn't realize *I was responsible* for ensuring the company met the FDA's quality systems requirements. On the product side, my experience with clinical diagnostics was as a patient. I knew a bit more about buoyant density separation because Alfa Wassermann had initially hired me as a consultant to identify new uses for this technology. At the time, single-cell proteomics was an emerging field, and different cell types have different buoyant densities. There was an opportunity for AWI to enter this potentially lucrative research market with a smaller automated instrument.[4]

---

2. Mel initially worked for Ascend Communications, which was acquired by Lucent Technologies (formerly Bell Labs) in 1999. The company later merged with Alcatel after a series of financial setbacks.
3. With the help of Alfa Wassermann S.p.A., the parent company, we developed a new, substantially pure form of the drug and a patented novel process to manufacture it. When I left Alfa Wassermann, all work on the project stopped. I then tried to license the rights directly from Roche, without success.
4. That product became the AW Promatix 1000.

We moved to the East Coast, and I gave Mel a tour of the company.

"We're going to launch a new benchtop clinical chemistry analyzer," I told Mel, and introduced him to the manufacturing director. "Sorin, can you give my husband a demonstration of how the new analyzer works?"

"Sure," he said, and put the machine through its paces. He showed Mel the printed output.

"This is great. How do you manage version control of the software?" Mel asked.

"Version control? We have a couple of programmers who make changes when something doesn't work," Sorin answered.

"There isn't version control?" Mel asked.

"Not really."

"Betsy, do not launch this product without version control of the software," Mel said. "It is the only way to keep track of what's been done."

The next day, I met with the regulatory manager, who confirmed that the software system was out of compliance. Based on her information and Mel's recommendation, I stopped the product's launch until the software system met the FDA's requirements.

The next pivotal moment came when a customer notified the company that the instrument was giving a faulty result on one clinical chemistry test. I contacted a nephrologist I'd worked with on CellCept to determine whether our error had put patients at risk. He assured me it hadn't. We corrected the problem and notified the FDA, which we were required to do. This triggered a three-day Quality Systems Inspection Technique audit. I learned more about the processes the company had and didn't have during those three days. Fortunately, the FDA inspector passed us unconditionally—the first time in the company's history.

The mostly absentee CEO was less appreciative upon learning about the many issues the audit revealed. He thought I should have already known about them and fixed what was lacking. Fair enough. I was grateful to be fired and moved back to California in time for the start of the school year.

People say the best time to look for a job is while you're employed. I've never been able to do that, because I am completely committed to doing my current job, even when it is no longer enjoyable or productive. But I do get fired frequently. I reach a point where I am bored and probably become a pain in the butt to my supervisor. I've been fired by 25 percent of my female bosses and 50 percent of my male bosses. Fortunately, I've been good at getting severance. I always stayed connected to the end and then negotiated a good separation once the writing was on the wall.

After the diagnostics company, I moved semi-seamlessly into consulting, with an emphasis on startup companies. First, I consulted for a venture capital firm, which didn't work out, and then I consulted in neuroscience with a company based in Heidelberg—which was good for my German and a wonderful place for Christmas shopping. I later co-founded a company based on optogenetics with a neuroscientist I'd worked with for four years and recognized when it was time to leave the company.

Today, I consult and help out where and when I can on projects I enjoy. These days, I do more writing than consulting, but I'm always ready to explore new opportunities and slide up the learning curve or even chase an old dream.

In the Path of the Forty-Niners

The Tevis Cup is a grueling hundred-mile endurance horse ride across the Sierra Nevada following trails made by fortune seekers 150 years ago. It had been a long-held dream of mine to compete and ride the hundred-mile trail in twenty-four hours. The race—a brutal test of both horse and rider—demanded six qualifying fifty-mile races before entry. Ever the data-driven planner, I mapped out my path to eligibility.

I trained on an endurance-tested Arabian near Berkeley, logging miles through the East Bay Regional Park's 1,200 miles of trails. But I needed more—I wanted to do it all in one year. So, in winter, I flew to Florida to train and compete with the top endurance couple. What followed was the most punishing weekend of my life—twelve-mile canters, relentless back-to-back races, and a bottle of Advil consumed in record time.

By March, I had five races under my belt but faced a problem: my friend's horse wasn't Tevis material. Enter Melissa and Robert, Tevis veterans who leased horses to qualified riders. They agreed to take me on—contingent on my surviving Robert's brutal training.

What followed were months of running, riding, and hiking in blistering heat, commuting hundreds of miles weekly to prepare. My sixth and final qualifier was a race on Saturday to be followed by another fifty-mile race on Sunday—a simulation of the Tevis race. The camp saying was "if you can complete two days at Cooley Ranch, Tevis is a breeze." I felt ready—until my feet betrayed me.

Melissa told me I'd be riding the entire race, so I wore clunky boots unfit for hiking. Robert had other plans. "Time to dismount," he announced. I protested. He insisted. I trudged up hill after hill, blistering my feet and losing my big toenail in the process. By mile forty-five, I was in agony, my feet dangling uselessly outside the stirrups. I finished the race—but the thought of another fifty miles the next day seemed impossible.

That night, I wrestled with a question: Was I willing to surrender control of my body to Robert's regimen? Tevis was doable—but only if I dictated when I walked and when I rode. By dawn, my decision was made. I found Melissa.

"I'm out," I told her. "But what if we rode the Tevis trail together over three days?"

She smiled. "That's a great idea."

And so, instead of a twenty-four-hour blur of exhaustion and stress, I actually saw the trail—the historic canyons, the abandoned gold mines, the forty-niners' gravestones in Deadwood. I dunked my head in the icy American River before cantering into the Auburn Fairgrounds, dizzy from heat and exhaustion but exhilarated.

I did Tevis my way. The trade-off? No silver buckle. But I didn't need one—I had the ride of a lifetime.

Marshall Matters

During these changes and disruptions, I had a child to consider. Marshall learned that life can come in two- or three-year increments. Imagine being in fifth grade and thinking you're settled in Europe when your mother suddenly announces you're moving back to California. It was difficult for him in some ways, but he made friends quickly and kept in touch with them wherever we were. Fortunately, he is a gracious, empathetic, lovely human being with tremendous emotional intelligence.

Wherever we were, I always hired bright and patient young women from the local universities to tutor Marshall. He was an adaptable child—whether because of nature or nurture, I don't know. When we told him he was adopted, he accepted it as just how life is—it wasn't a big deal for him.

He was about five years old when we explained how he became part of our family. Mel and I had waited until we thought Marshall was at an age when he could understand. We always made his needs a priority and made a considerable effort to provide the best environment we could. Nevertheless,

I sometimes wondered whether we were the right parents for him. Mel and I adhered to our rule that one parent would always be on the same continent with him. I know this sounds absurd, but we stuck to it. Once, Mel made a three-day turnaround from California to Switzerland when I had a business trip to the US.

My mother used to say that it doesn't matter who provides emotional support for a child during the first two years—only that the child has it. Marshall had support from Mel and me, my mother, and Tina, our loving Chilean housekeeper.

Throughout the years, Mel kept in touch with Marshall's birth mother, Patty, and sent her updates. His birth father, Jim, whom Patty had married, remained disinterested.

Soon after Marshall was born, my mother asked Patty if she would be a surrogate mother for Mel and me. She declined. That was typical of Mom. She liked Patty, so why not ask her if she would mind having Mel's baby? Until Marshall was eighteen, Patty kept him a secret from everyone but her father and Jim. We had told Marshall about Patty's son and Jim's daughter—his half-siblings—but that had not been reciprocated.

Patty had hoped to attend Marshall's high school graduation. At that point, she confessed to the rest of her family that she'd given birth to two more children, one of whom was Marshall. Unfortunately, she got sick before Marshall's graduation and was unable to attend. Her disclosures led to Jodi, Marshall's half sister, contacting him on Facebook. They became good friends.

After high school, Marshall spent a month with Jim and Patty at their home in Iowa. We decided a month was long enough for everyone to get to know each other. We told Marshall he could stay longer if he felt more comfortable there but with one condition: He needed to come home for Christmas.

I wanted to do whatever Marshall wanted. If he wanted to stay in Iowa with Jim and Patty, that would be fine because it was his choice. Would we be hurt and disappointed? Yes. Would we want visiting privileges? Absolutely. I was open to the possibility that we might be all wrong for him in the same way that biological families can be wrong for each other. We faced it.

Three days before the month's end, he called and asked whether he could return as planned. "Of course" was our unambiguous and unanimous answer.

He'd needed to see if they were a better fit, but he came back to us and was happy to be home.

Lessons Learned

- Listen, learn, move on when it is time, and get severance!
- Get products right the first time, even if doing so appears to take longer.
- Commit 100 percent, but don't be afraid to adjust the goal and enjoy the journey.
- Be straight with what you know and what you think; don't color the facts to placate who you're talking to—causing ripples is okay.
- You need to listen to what the environment is telling you, what is working for you, and what is not working for you.
- Don't bother trying to control others.

Chapter Thirty-Two:

How Incredibly Strange to Be Seventy

As part of writing this book, I posted some videos on YouTube so my friends and family could see what life had been like on Malo. I'd filmed the movies in 1975, and they conveyed a sense of the environment and people.[1] Two years later, in 2019, a comment appeared on one of the videos: "Bob, did you upload these? Love them."

It was from Sumbue, Jole's son. We exchanged emails, and I relayed that Bob had died after a successful career in geriatric anthropology. Sumbue, now an ambassador for Vanuatu, was living in Geneva. I would be in Europe that summer and planned to visit friends in Switzerland, so of course I'd visit him. He asked whether I had any photos of Malo and, if so, could I bring them. I didn't think I did, but to be sure, I rummaged through our basement and opened boxes that had lain dormant for forty years.

Inside were hundreds of slides and negatives. I had them digitized, and I also purchased copies of my and Bob's dissertations. On the train from Basel to Geneva, I reviewed fifteen hundred photos spanning the years 1973 to 1982. It was a rush of memories. That night, I met Sumbue and his wife, Rose, at their apartment. Together we spent hours catching up on nearly thirty years of family, friends, and history. We also exchanged gifts. I brought a bottle of Rioja wine and fresh dark chocolate truffles from Basel, and they gave me a magnificent Mother Hubbard. As always, I took copious notes, this time in a Snoopy notebook I'd purchased at a corner store on my way to meet them.

1. A video of wedding preparations is available here: "Wedding Malo 1975," posted September 1, 2017, by malovanuatu, YouTube, https://www.youtube.com/watch?v=7cOIAcx9zGQ. A video depicting Avunatari customs is available here: "Avunatari 1975 Bob Custom," posted September 1, 2017, by malovanuatu, YouTube, https://www.youtube.com/watch?v=EHhU-4cDGAw.

They told me that Jole and Vomule were dead. Sumbue showed me a picture of Vomule that he had taken from a frame of my YouTube movie.

"My mother was beautiful. I never knew she was so beautiful."

It brings tears to my eyes as I write this—I remember the funny, spirited, intelligent, practical woman who helped me in subtle ways.

The next morning, we met for breakfast.

"Betsy, I was up until 4:00 a.m. looking at the photos. It was wonderful to see old man Muele leading a custom dance. There were photos of people I didn't know or didn't recognize, but so many pictures of people I knew."

I was amazed that he remembered so much, because he had been a young boy then.

FIGURE 58: Sumbue in 1975. On Malo during my second visit (about nine or ten years old). (See also Color Plate 11: Sumbue and Rosemary Antas [Geneva, 2019].)

I left Geneva that day, but a week later, I wrote and asked him what it would be like if I visited Malo again. He wrote back that people were excited by the possibility of a visit. I also had heard from Heather's daughter that her father, Ted, was not doing well. At that moment, I realized it was much better to see people while they were alive than attend their funeral or memorial. I had a twelve-day window in September 2019 and bit the bullet.

Back into the Wild Blue Yonder

My trip to Malo was carefully crafted. I had a quick stop in Hawaii to see my former assistant, Sabine, and revisit Makapu'u, the site of my near-fatal body surfing adventure. I flew direct from there to Brisbane and spent thirty-six hours with Ted, Heather, and their family. I am grateful for that time because Ted died two months later.

My next stop was Port Vila to rendezvous with Carol and Brian's son—the same Carol who fixed me up with Mel.

Port Vila has never been special for me, but I was able to donate the photos and dissertations to the Archives and the Vanuatu Cultural Centre. Sumbue texted that his sons would meet me at the airport. Struan and his daughter were there, but Stanleyson[2] (of Stan and the Earth Force, Vanuatu's popular band) was busy recording. Sumbue's youngest son, Tavuti, was "coincidentally" going to Malo, but he was really there as my escort. We flew to Espiritu Santo together in a Boeing 737 and shared a taxi into town.

Santo—now known as Luganville—was very recognizable. Many of the buildings had been repurposed, but much was familiar. I left my luggage at the same Hotel Santo as where I'd previously stayed. Shouldering a new backpack, I met Tavuti. The backpack was loaded with supplies for the two days I'd be on Malo, including antimalarial medication because Malo and Santo are bright red on the CDC's malaria map. I also brought special tropical mosquito repellent, a mosquito net, bottled water, and peanut butter—just in case it was the only thing I could eat. There was a guest house on Malo that looked nice, and I had that in my mind as plan B. But I didn't want to offend anyone and figured I could sleep on a bamboo frame if need be. I didn't know what to expect. (See Color Plate 12: Outdoor market in Santo [Luganville, 2019], and Color Plate 13: fruit bats [flying foxes].)

We collected relatives in Luganville for the trip to Malo; they were Jole and Vomule's grandchildren and great-grandchildren. By the time we were ready, it was well after five o'clock, and darkness comes quickly in the tropics. I suggested maybe going the next day, but Tavuti insisted that we were going now. At the shore, Alan, Sumbue's youngest brother, was there with an open-hulled fiberglass boat. The wind was blowing, there were whitecaps, and we

2. His song "Equal Opportunity" gives a fair picture of what life was like on Malo in the 1970s; it is available here: "Equal Opportunity," posted March 2, 2015, by thepacmasvoice, YouTube, https://www.youtube.com/watch?v=V_jr-vZugYc.

were going to cross the open sea. With lots of trepidation, I got in the boat, telling myself that Alan did this all the time and he would not risk the lives of so many family members with a dangerous crossing. Despite telling myself that, I was relieved when we pulled in to the Avunatari beach and made it to my old home site just as dark descended.

So much seemed the same. I'd have easily found my way "home" on the trail. Dinner was traditional *laplap*, cooked in a stone hearth. Water was boiled in the same type of aluminum kettle that had been used decades before, and we talked for a good long time about people and changes while we sat around the kitchen table (men on one side and we women on the other). I stayed in Struan's house. To my relief, there was both a mattress and a mosquito net.

What had changed? The road around the island had expanded to two lanes. Solar power had also come to Malo. Tiny panels powered smartphones and lights. In addition, the water supply was sealed and stored in tanks that held many hundreds of gallons.[3] These were huge differences.

FIGURE 59: A surfaced two-lane road. No more single-lane dirt road around the island.

3. The Red Cross put in numerous water tanks in 2018. It was a huge step in water safety and likely in malaria control. "Over 3,000 Villagers in Santo Rural Benefit from Rainwater Tanks," Vanuatu Red Cross Society, November 5, 2018, http://www.vanuaturedcross.org/news/2018/11/5/over-3000-villagers-in-santo-rural-benefit-from-rainwater-tanks-nbsp.

FIGURE 60: Elegantly landscaped and maintained properties

FIGURE 61: Solar panels for lighting

FIGURE 62: Water-collection tanks

There was an outhouse at Struan's place and a separate shower with a bucket. The compound also had numerous individual houses with a common kitchen hut and shared bathroom facilities. All the compounds I saw were immaculately clean and tidy. Every place I visited had sealed water and was meticulously maintained. Remarkably, I never heard the buzz of a mosquito or any other insect.

FIGURE 63: Avnaro then (1975)

FIGURE 64: Avnaro now (2019). Home to Jole and Vomule's descendants.

On Sunday, I went to church. Instead of the sermon, songs, and readings being in Tamambo, as they had been in the 1970s, everything was in Bislama. Additionally, the Bible passages no longer referred to places in the Middle East but instead to Vanuatu. Pastor Seth asked me to address the congregation. In Bislama, I said what a pleasure it was to be there and thanked everyone for their generosity during my many visits. Afterward, I joined Pastor Seth outside, where we shook hands with each of the nearly one hundred parishioners.

"You took blood belong me time you stop long Malo,"[4] person after person said. "Me one small girl no more." These were middle-aged people who remembered me taking their blood when they were children.

There was a community feast afterward, and I talked to everyone, including Vomoli Havo, who had been my house girl and who was still very recognizable. We walked to Malahatou to visit Voligo, her mother, who was now seventy-eight, crippled and housebound, waiting to die. Vira had passed a few years earlier. I sat next to Voligo. We hugged, and both of us had tears in our eyes. For the next hour, we talked nonstop. Her warmth was palpable and our bond steadfast. So many years had passed and yet we were very much the same.

That evening, Tavuti, a budding journalist, interviewed me at Sumbue's request. Alan and I talked about my research, and I bonded with five-year-old Lela, who brought me *ngai-ngai* nuts. The next day, we visited the other person I knew from "the old days." Sangavulu was now eighty-nine and used a wheelchair. I promised to send a copy of the photos and dissertations to him through his grandson, who worked in the airport in Port Vila.

4. A rough translation is "You took my blood when you were here on Malo a long time ago."

FIGURE 65: Visit with Voligo (2019). Spending time with her, her daughter Vomoli Havo, and her grandson.

FIGURE 66: Visit with Sangavulu. An ever-present business leader who ran the Malo co-op.

The SanMa[5] Day celebration was underway on Monday. I stopped by briefly. The social structure was still the same as I remembered, with men on one side and women on the other. I sat with the women, and then some of the men beckoned for me to join them. I did—nothing had changed. I returned to spend more time with Voligo and gave her the Extra-Strength Tylenol I carry as a precaution. We hugged, knowing we would not see each other again and grateful for our reunion. I walked to the reef to take the boat back to Luganville. Sumbue's younger brothers, Alan and Donald, and their families came to see me off. Lela was in tears. This crossing to Luganville was even more perilous than the first.

I returned to the Hotel Santo wearing my Mother Hubbard. On the way to my room, I passed a couple who were sitting on their patio. The man was wearing a T-shirt with the words "California Surf" on it. I took the bait and asked whether he was from California.

In a clear Aussie accent, the man said, "No, but from the sound of it, you might be."

We chatted and his partner, Jenny, asked, "What brings you to Luganville?"

"I did my doctoral research in biomedical anthropology here in the 1970s. I had the time and I wanted to see what it was like now."

"Biomedical anthropology? What were you studying?" she asked.

---

5. SanMa is the name of the northern province in Vanuatu and combines the names of the two largest islands, Santo and Malo. SanMa Day is September 23.

FIGURE 67: Saying goodbye. At the reef with the Antas family members before leaving Malo.

"The natural history of infection with HBV."

"Sit down!" she said immediately.

That began our hours-long discussion. Jenny was a physician and Paul was a builder. Together, they had been coming to Vanuatu for twelve years and building A-Posts (dispensaries). I gave her a copy of Bob's and my dissertations. Later, I received an email from her:

> *We both also enjoyed hanging out with you in Luganville . . . It was a very important learning experience that not all gray-haired women in Mother Hubbard dresses are missionaries, and they can quickly transform into kava-drinking buddies who have researched subjects which I have pondered for years. Your thesis on HBV was fascinating, hard to believe that it is over forty years old. It raised so many interesting questions (e.g., differences in location for birthing) which are still relevant today. It was really useful to read about usual post-natal practices in your dissertation. I have also been discussing it with my medical anthropology buddies.*

I rented a car and visited Champagne Beach, where Mel and I had swum during our honeymoon. On my return to Vila, I splurged and overnighted at Le Lagon, where I'd spent the days leading up to Christmas in 1975 and where we'd stayed during our honeymoon. At the airport, before I checked in, I asked a man if he knew Gilman Imbulu. Everyone around me started to laugh. That man was Gilman—Sangavulu's grandson. I gave him the USB stick with the photos and dissertations I'd promised.

The connections were everywhere.

I hope to return one more time (at a minimum) with Marshall and Mel. I want Marshall to have a visual memory of this experience and for Mel to appreciate the changes. Since my return, I've learned that 90 percent of Malo was destroyed by Hurricane Harold. I was fortunate to have seen it in its prime.

Lessons Learned

- You can always return to where you've been.
- Pay back what you can with what you have saved.
- People can remember what you did or said for decades.
- Keep taking opportunities to meet new people.
- It is better to see people while they are alive than to celebrate their life when they're gone.
- Do things while you can—everything changes.

Chapter Thirty-Three:

Malo—Reflections and Reminders

The pre-Malo Betsy was very different from the post-Malo Betsy, and those differences shaped my life and career. I am an ordinary person, but I have had the courage to choose opportunities that became extraordinary. An important choice that I made was to study anthropology. This decision was the gateway to analyzing Polynesian languages, which led to my fieldwork in Melanesia. Despite my training, I made mistakes and overcame obstacles. Success didn't shape me—my mistakes did. I faced my mistakes, changed my behavior, and moved forward with new insights.

In this iterative process of trial and error, I pushed the pendulum out of accepted behavioral norms or intellectual dogma and was recognized both as gutsy and creative as well as outrageous and disruptive. Listening, feedback, and real-time adjustments are critical to effective risk-taking and learning.

My blueprint for jobs has generally worked. I won't accept a position if it doesn't feel like play or doesn't offer something new that makes it exciting. If those elements are missing, then that job is not for me. I need to have the freedom to explore. My best work happens when the subject matter is new and interesting or the problem seems intractable. These moments are invigorating, but they are sometimes frustrating until either a synthesis or a solution emerges. The more skills and experiences one has, the easier it is to find novel solutions or discover important, actionable connections.

Fieldwork and the subsequent data analysis made it clear to me that my career would be in a dynamic field where the information constantly changes and the problems are complex. Branding and pharmaceuticals sated my need for "mind candy." My search for constant mental challenge pushed me to move horizontally into new areas rather than seeking promotions by doing the same job but with a more prestigious title. I took each step of my career to learn new skills or competencies and explore the unknown. I was fortunate and worked hard to have this luxury.

Did the Malo Islanders have a long-lasting effect on me? I still peel fruit "their" way by scoring an orange or grapefruit skin before peeling it. I cut fruit "their" way, into four unequal-sized pieces around the seed core. If I'm stuck carrying a heavy shoulder bag, I will, on occasion—if I'm not too intimidated by people gawking at me—carry the bag with the strap around my forehead so the bag rests on my back, the same way the islanders carried their yam-laden baskets from the garden. It is an easy and efficient way to transport a heavy load and have your hands completely free. I bundled my son, Marshall, in a cloth sling on my hip until he was a toddler. This was much easier for me and more comfortable for him than stuffing him into a baby pack. I also premasticated his food, which kept my life simple and gave him a great palate.

Most importantly, the islanders taught me that if I am satisfied with what I am doing and it's interesting, that can be enough. I learned to adjust my goals based on real-time information, and I discovered that taking opportunities to explore different roads can lead to extraordinary experiences. Buckle up!

Epilogue

Life always has more in store. Since I finished writing this book, Marshall completed his college degree and has worked in Hollywood media productions. He married his longtime Greek girlfriend, Argyri. She is a rising talent in the field of digital law and intellectual property. Presently, they have a home in Canada and maintain an apartment in Los Angeles. Marshall dreams of writing for scripted television.

FIGURE 68: Marshall starting his career

Mel and I still live in Berkeley in the house we bought in 1984. I have sustained close, strong friendships from every phase of my life. At my seventieth birthday party, there were friends I've had for more than sixty years, forty years, and twenty years. It's been good, and I'm grateful.

And I know there's more to come.

If you enjoyed the book, please post a review on the site where you purchased it. You are welcome to visit my website at www.ElizabethReedAden.com to learn more about Vanuatu. My best-selling and award-winning medical thriller *The Goldilocks Genome* uses the epidemiology techniques I learned to solve a pharmacology mystery. For those interested in learning about the Goldilocks effect between their prescription drugs and their genes, my free download Drugs & Genes is available on my website. In addition, my website includes free information on personalized medicine. I also published and republished the award-winning vintage cozy mysteries written by my godmother, Eunice Mays Boyd, between 1943 and 1970.

To learn more about Elizabeth and *Mud, Microbes, and Medicine*, visit her website or scan the QR code below.

Appendix I:

The Geography

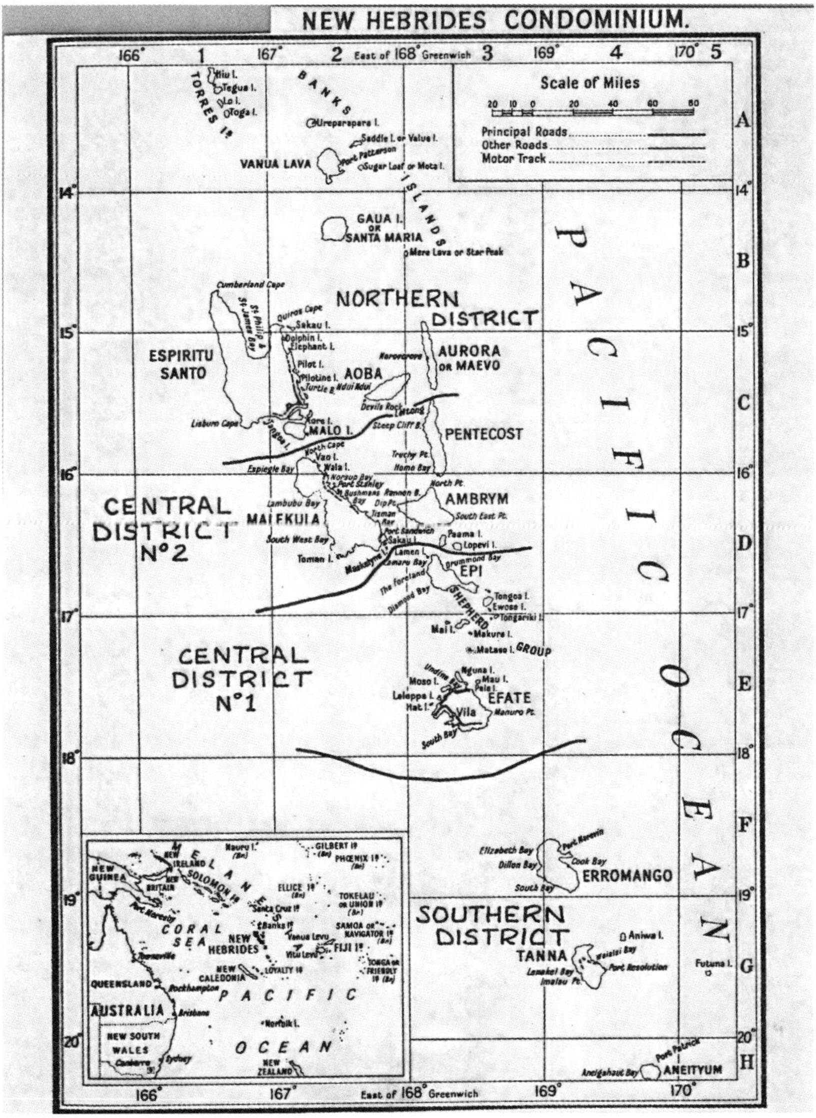

FIGURE 69: Administrative map of New Hebrides (1976)

Appendix II:

HBV Epidemiology and Biology

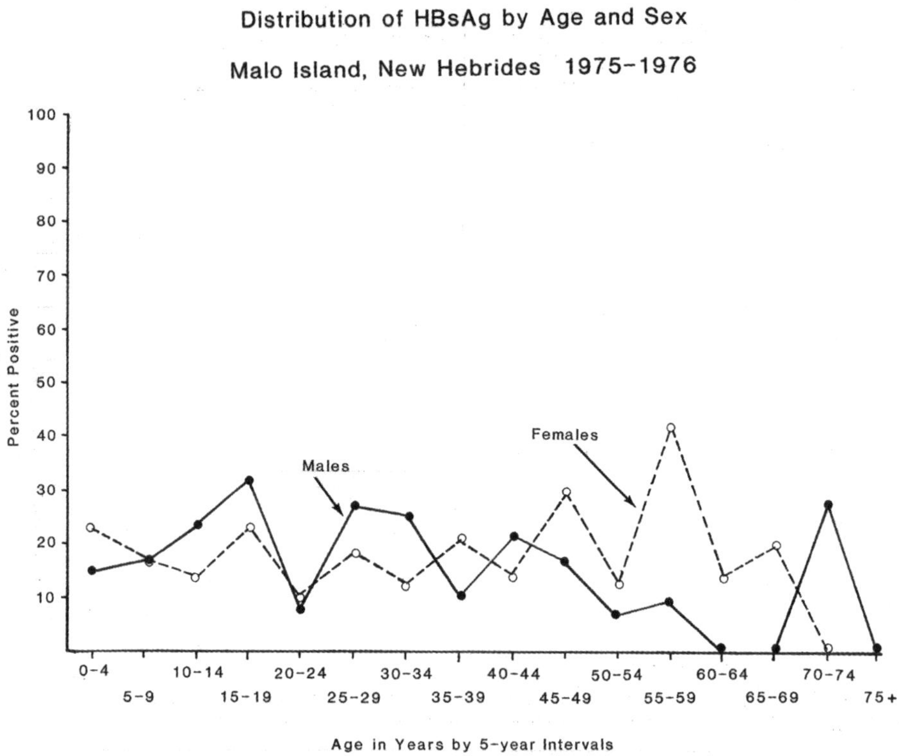

FIGURE 70: Prevalence of HBsAg on Malo (1975–1976)

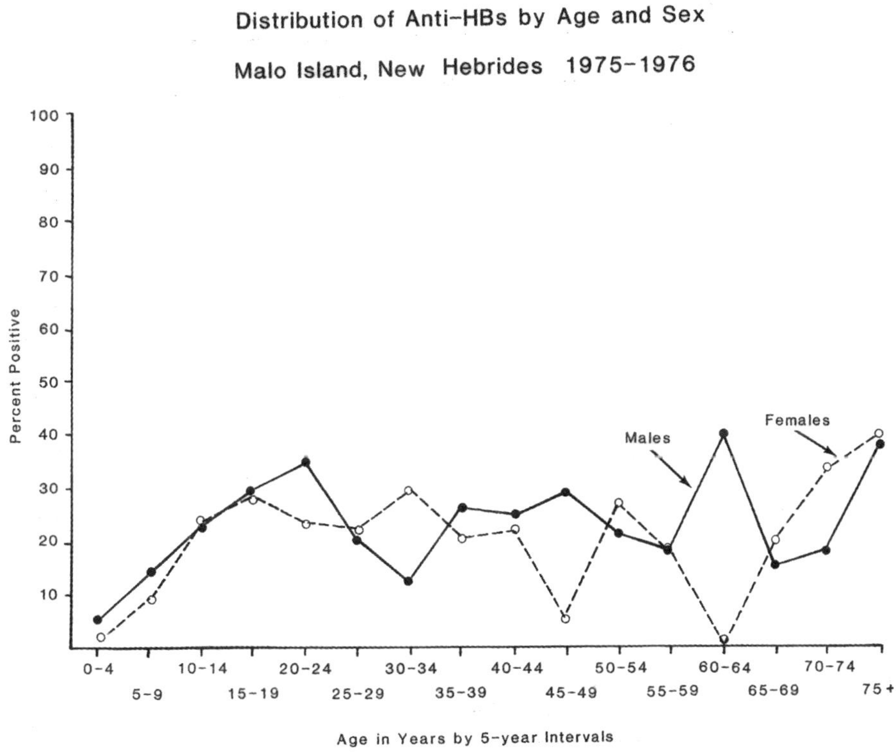

FIGURE 71: Prevalence of anti-HBs on Malo (1975–1976)

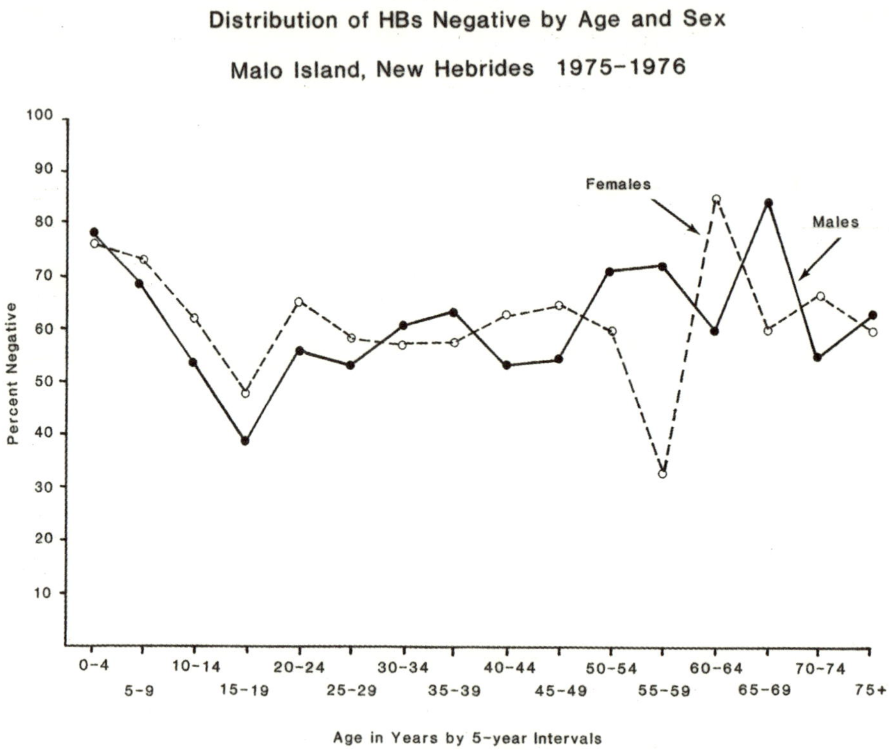

FIGURE 72: Prevalence of negative for HBsAg on Malo (1975–1976)

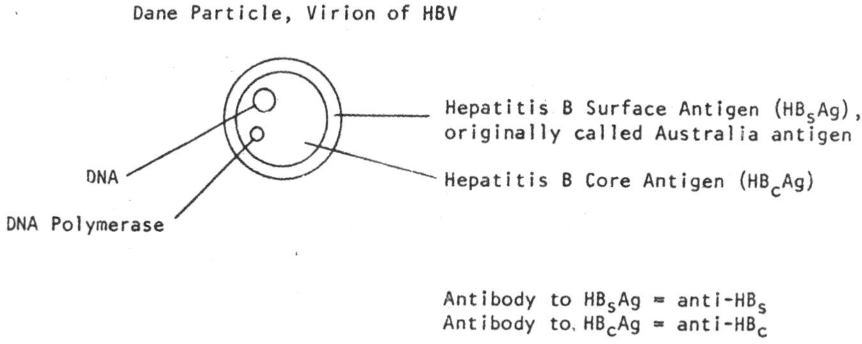

FIGURE 73: Structure of the hepatitis B virus

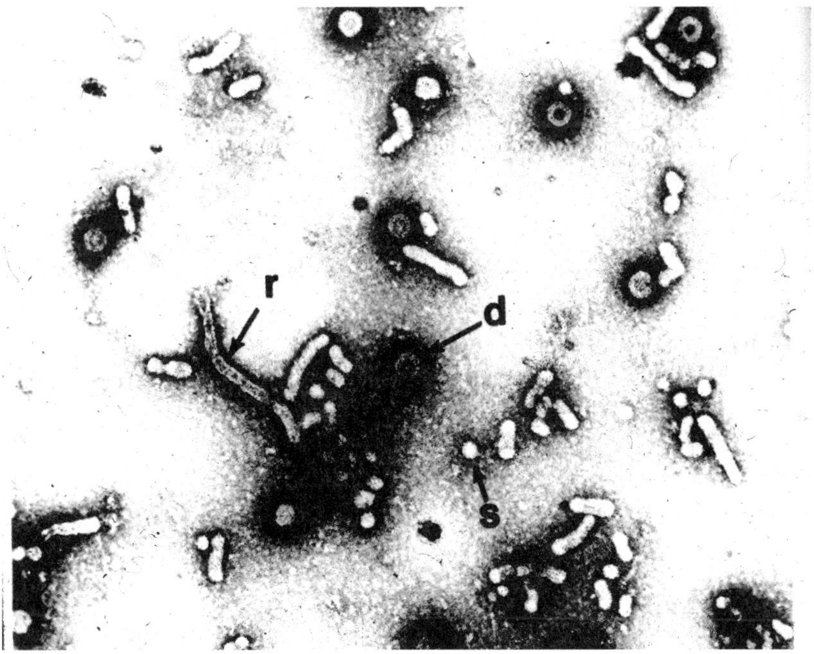

FIGURE 74: Electron micrograph of HBV. (d = Dane particle; s = spherical form of surface antigen; r = rod form of surface antigen).

Appendix III:

Epidemiology Observation Formats: Written and Checklist

Louise June 24, 1976 pg. 2

:40 M looks away and the I sucks and the I moves its hand around and
the M has her hand on the breast and the I hits the breast and the M
looks to the I and the I rubs the Is chest and then M teat comes out
and then she puts it in her dress and then M sits the I up and it is
standing on her leg and facing the O and not the M and the M smiles
and the I voc.
:41 M has her hand under its butt and her hand on its chest and
then lifts the I up and then she looks to the I and brushes its shoulder
and then sits the I on her lap facing her and the I looks away and the
M looks to the I and the M is holding the Is hands and she stands
:42 the I up on her leg and bounces the I up and down and holds the I
upright and bounces and the I looks to F and M bounces the I toward
:43 F and the M smiles to F and the I looks to F and I drools and bounces
:43 M sits the I down on her lap and the M looks to I and then M wipes
the Is mouth and xxxxxsways the I and the I looks away and moves its
body and voc and M holds its arm and pats its side and the I frets and
the M stands the I up and the M looks to I
:44 M bounces the I and M looks away and bounces the I up and down and
M looks to F and then to I and the I looks away and voc and M looks to
I and the M turns the I to face her and clucks to it and smiles and
the I fusses and the M bounces the I and holds it up to her mouth and
the I frets and
:45 M holds the I away and then bounces and then lifts the I up to the
xxxxxxxxxxxxxxxxxxxxxxxxxx and then down to nursing position on the
same breast and xxxxxxx M looks to I and the teat is in its mouth and
the M jiggles and then sways and then jiggles and the I moves its hand
all over and the M looks away and makes a big jiggle and then stops and
then M brushes the Is leg and jiggles it and then looks away

FIGURE 75: Ethnographic observations. Typed notes transcribed from initial observation period.

Disease Transmission – Observational Checklist

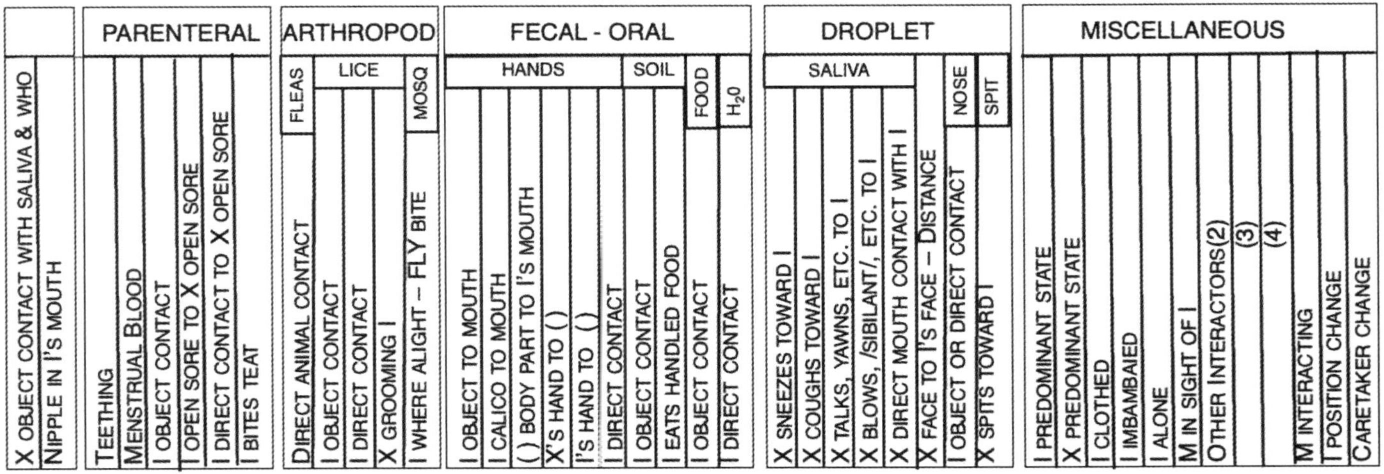

FORMAT
CHECKS MADE IN BOX(ES)
20 SECOND INTERVALS
ABBREVIATIONS/NOTES FOR INTERACTORS

LEGEND
I = INFANT
M = MOTHER
X = OTHER INTERACTOR (DEFINED)
( ) ABBREVIATIONS USED FOR BODY PARTS

OBSERVATION TIMES
9:00-10:00
10:30-11:30
13:00-14:00
14:30-15:30

FIGURE 76: Ethology observations. Observation checklist based on transmission.

Appendix IV:

Interactors with Infant by Stage of Development

Stages of Ni-Tamambo Infant Development
(0-52 Weeks)

Stage I (*curanji daeca* "red child")
Approximately 0-6 weeks
Focuses on on objects
Muscular control of the limbs
Earliest hand-to-mouth activity
Partial head movement control

Stage II (*curanji malalum* "soft child")
Approximately 7-20 weeks
Complete support of head
When held upright:
- moves feet up and down,
- bends knees

Sits supported and
Sits briefly when unsupported
Stands unsupported briefly

Stage III (*curanji woriwori* "small child")
Approximately 21-46 weeks
Sits unassisted
Hand-to-object activity
Object-to-mouth activity
Forward rocking motion to knees while sitting
Squatting
Sit-crawl (optional)
Crawl (bottom of feet, not on knees)
Stands alone from squat
Holds stand position briefly
Pulls self up to a standing position
Walks a few steps

Stage IV (*curanji suica* "strong child")
Approximately 46-52 weeks)
Walks strongly unassisted

FIGURE 77: Infant developmental milestones by stage

Tamambo Infant Development

Stage I

Primary Caretaker(s):	Mother
Secondary Caretaker(s):	Father Father's mother
Primary Interactor(s):	Mother

Predominant Modes of Transmission
Arthropod: Flies and/or Mosquitoes
(Not specific to an individual)

Tamambo Infant Development

Stage II

Sources of Infection		Mode(s) of Transmission
Primary Caretaker(s):	Mother	Droplet
Secondary Caretaker(s):	Father Eldest Daughter	Droplet Droplet
Primary Interactor(s):	Mother	Droplet

Tamambo Infant Development

Stage III

Sources of Infection		Mode(s) of Transmission
Primary Caretaker(s):	Mother	Fecal-oral Arthropod (Head lice)
Secondary Caretaker(s):	Eldest Daughter	Fecal-oral
Primary Interactor(s):	Adjacent Siblings	Droplet

FIGURE 78: Stage of infant development and primary interactors

Appendix V:

Serology and Epidemiology Results—Development of Chronic Carriers

DEVELOPMENT OF THE HB_SA_g CARRIER STATE

MALO ISLAND

SOURCE OF EXPOSURE	INFANT: # INFECTED WITH HBV < 1 YEAR		INFANT: # HBsAg CARRIER 2-3 YEARS	
	# INFANTS	# FAMILIES	# INFANTS	# FAMILIES
MOTHER +	7	10	6	6
ONLY FATHER +	3	3	0	2
ONLY SIBLING +	1	2	2	2
NO RELATIVE +	1	11	0	15
TOTAL HBsAg(+)	12	26 (46.2%)	8	25 (32.0%)

FIGURE 79: Overview of infant HBV carrier status by family-member HBsAg status

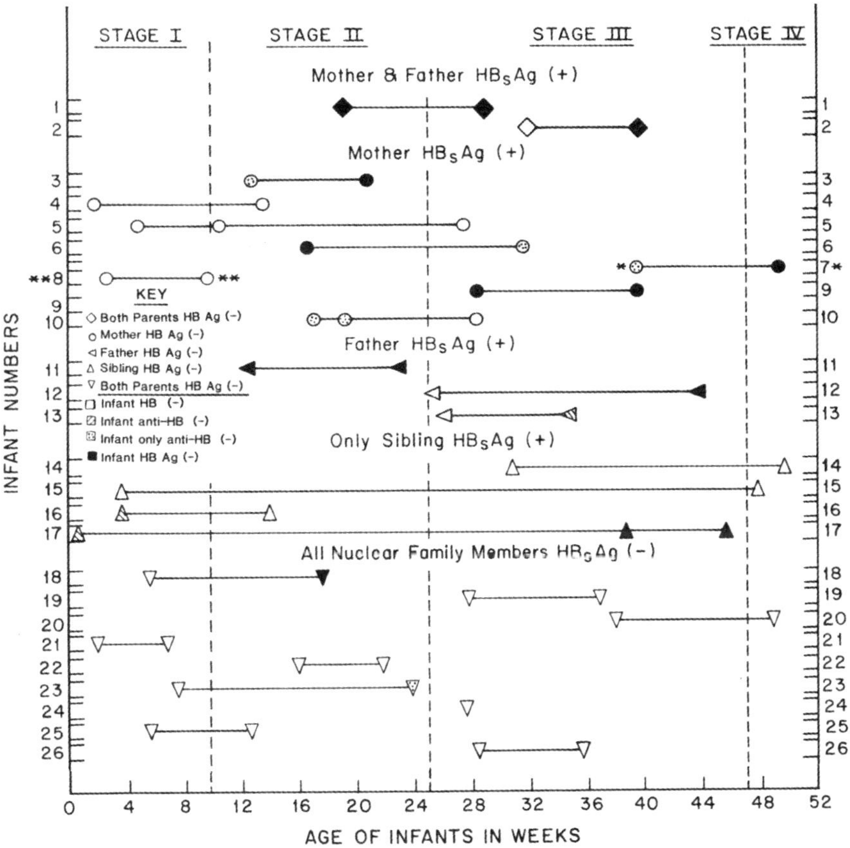

FIGURE 80: Infant HBsAg status by stage at time of observation (1976)

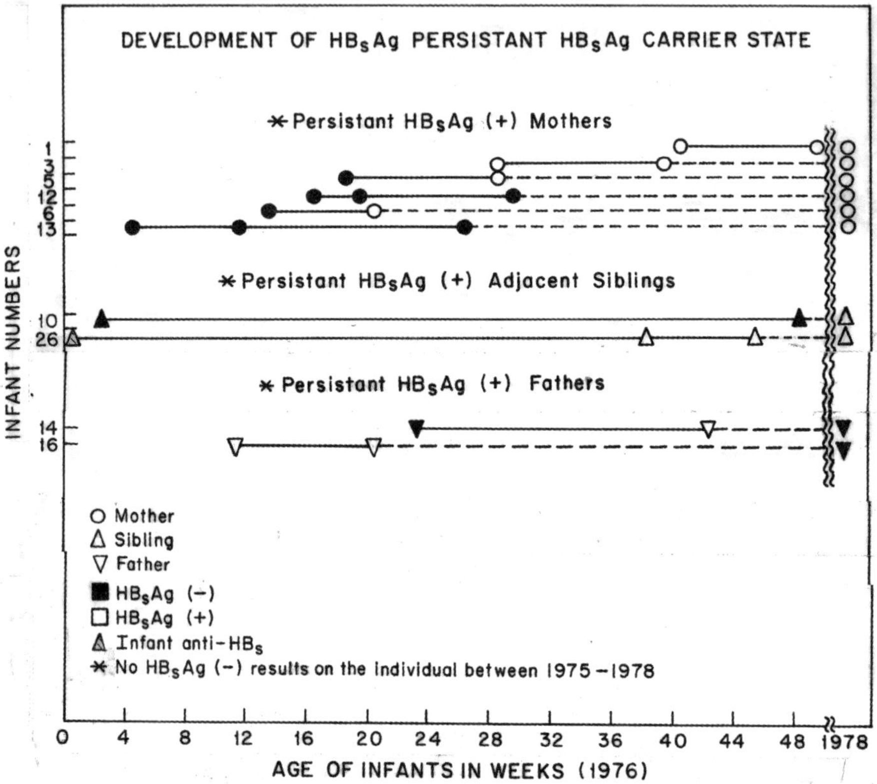

FIGURE 81: Infant HBsAg chronic carrier status (1978)

Bibliography

Alter, Harvey J. and Baruch S. Blumberg. "Further Studies on a 'New' Human Isoprecipitin System (Australia Antigen)." *Blood 3 (1996): 297–309.*

Beasant, John. *The Santo Rebellion: An Imperial Reckoning.* Honolulu: University of Hawaii Press, 1984.

Biddulph, Jim. "Squabbles in the South Pacific: The British and French Will Leave No Legacy of Cohesive Government to the New Hebrides." *The Listener*, June 26, 1980: 816.

Blumberg, B. S., B. Larouzé, W. T. London, B. Werner, J. E. Hesser, G. Saimot, et al. "The Relation of Infection with the Hepatitis B Agent to Primary Hepatocellular Carcinoma." *American Journal of Pathology* 81, no. 3 (1975): 669–82.

Dickie, Elizabeth Reed. "Family Behavior and the Transmission of Hepatitis B Virus in Malo, New Hebrides: Ethology, Ethnography, and Epidemiology in the Study of the Natural History of Disease." PhD diss., University of Pennsylvania, 1979.

Dickie, Elizabeth Reed. "A Lexicostatistical Analysis of Six Polynesian Languages: A Study in Methods." Master's thesis, University of Cincinnati, 1976.

Dickie, Elizabeth Reed. "A Note on the Naming of Malo Island, Vanuatu." *Journal of the Polynesian Society* 90, no. 1 (1981): 111–116.

Feitelson, M. "Hepatitis B Virus Infection and Primary Hepatocellular Carcinoma," *Clinical Microbiology Reviews* 5, no. 3 (July 1992): 275–301.

Greensboro Daily News. "Bow-And-Arrow Rebellion Splits South Pacific Islands." May 31, 1980.

Hedrick, John D. "Lapita Style Pottery from Malo Island." *Journal of the Polynesian Society* 80, no. 1 (March 1971): 5–19.

Hickey, Michael. "Why is Hemophilia Called the Royal Disease?" *HemAware—The Bleeding Disorders Magazine*, April 28, 2023, https://hemaware.org/bleeding-disorders-z/royal-disease.

IndexMundi. "Vanuatu—Life Expectancy at Birth, Total (Years)." Accessed June 9, 2025. https://www.indexmundi.com/facts/vanuatu/indicator/SP.DYN.LE00.IN.

Lévi-Strauss, Claude. *The Elementary Structures of Kinship*. Boston: Beacon Press, 1969.

Maedola, Fred. "Dollar Nomore, Makem Me Loosem Home." YouTube. June 17, 2021. https://www.youtube.com/watch?v=0KYJ7NSUo0M.

Malone, W. Scott. "Dilemmas of War—The Media." Black Box Intel Center. May 5, 2012. https://blackboxintelcenter.blogspot.com/2012/05/w-scott-malone-dilemmmas-of-war-media.html.

Malovanuatu. "Avunatari 1975 Bob Custom." YouTube. September 1, 2017. https://www.youtube.com/watch?v=EHhU-4cDGAw.

———. "Wedding Malo 1975." YouTube. September 1, 2017. https://www.youtube.com/watch?v=7cOIAcx9zGQ.

McIntyre, James K. "Investigations into the Relative Abundance of Intersexual Pigs in the Republic of Vanuatu." *Science in New Guinea* 22, no. 3 (1997). Reprinted in *The Southwest Pacific Research Project* (2016), https://www.swprp.org/intersexual-pigs.

Michener, James. "Return to the South Pacific." *Los Angeles Times*. October 19, 1986. https://www.latimes.com/archives/la-xpm-1986-10-19-tm-5883-story.html.

The New York Times. "New Hebrides, in South Pacific, Is Becoming Popular Tax Haven." May 10, 1971.

Roy, S., F. Stanczyk, D. R. Mishell Jr., M. Lumkin, and E. Gentzschein. "Clinical End Endocrinologic Study of Continuous Levonorgestrel Administration from Subcutaneous Solid Polydimethylsiloxane Rods." *Contraception* 21(6): June 1980: 595–615.

Rubinstein, R. L. "Placing the Self on Malo: An Account of the Culture of Malo Island, New Hebrides." PhD diss., Bryn Mawr College, 1979. Microfilm.

Sydney Morning Herald. "Andrew Stuart: Diplomat Who Fought 'Coconut War' for Vanuatu Independence." February 16, 2014.

The Telegraph. "Andrew Stuart—Obituary." February 7, 2014. https://www.telegraph.co.uk/news/obituaries/10625216/Andrew-Stuart-obituary.html.

Thepacmasvoice. "Equal Opportunity." YouTube. March 2, 2015. https://www.youtube.com/watch?v=V_jr-vZugYc.

Vanuatu Red Cross Society. "Over 3,000 Villagers in Santo Rural Benefit from Rainwater Tanks." November 5, 2018. http://www.vanuaturedcross.org/

news/2018/11/5/over-3000-villagers-in-santo-rural-benefit-from-rainwater-tanks-nbsp.

Wilson, L. G. "The Crime of Saving Lives: The FDA, John Najarian, and Minnesota ALG." *Archives of Surgery* 130, no. 10 (1995): 1035–1039.

Yumi TokTok Stret. "Vanuatu History Documentary (*YUmi Winnem Freedom*)." YouTube. June 24, 2015. https://www.youtube.com/watch?v=m8VfxQpvEKA&t=3s.

Afterword

The seroprevalence studies conducted by Dr. Aden showed the Malo Islanders had sky-high rates of infection, up to almost 25 percent in some areas. Disentangling vertical (mother-to-infant) transmission from horizontal (child-to-child) transmission was the subject of her dissertation. At the time, this was absolutely cutting edge, and her mentor, Baruch Blumberg, won the Nobel Prize in Physiology or Medicine for his work on HBV in 1976. Even now, with extensive vaccination and maternal screening campaigns, hepatitis B remains the most important cancer-causing viral infection we know of, and hepatocellular carcinoma—the consequence of chronic HBV infection—accounts for more than 850,000 new cases of hepatocellular carcinoma annually and 750,000 deaths. It is, in fact, the third most common cause of cancer deaths worldwide behind lung and colorectal cancers.

—George W. Rutherford, MD, MA, professor of epidemiology, preventive medicine, pediatrics, and history at the University of California, San Francisco

Acknowledgments

I want to thank Ewa Wojner, Dawn Johnson, and Mel Aden, who read and shared their comments and suggestions on the numerous versions. My editors, Linda Watson-Brown, Alan Rinzler, and Michael Denneny, provided valuable advice on structure. Many thanks to Mary Hediger, Caroline Stuckart, Jean Blumberg, Rebecca Ingraham, and Jenny Blackett-Smith for permission to use their letters and to Charles M. "Mel" Aden for allowing me to use his diary. Ward Goodenough's letters were reviewed by his children and used with their knowledge and without objection. Rights to use letters from Marilyn Reed Lucia and Salvatore Pablo Lucia were included in their estate. Many thanks to Anne and Jane Blumberg for their help as well.

I wish to thank the many people who helped during the final stages of the book, including Geoffrey Berwind, Michelle Orwin, Anne Durette, Dawn Johnson, and Ewa Wojnar. Fortunately, Patricia Marion and Barbara Werner reviewed the science.

To the best of my knowledge, the events are true and happened as described. The perspective is mine, and mine alone.

About the Author

Elizabeth Reed Aden is a biomedical anthropologist turned biotech/pharma strategist with a career spanning epidemiology, virology, and global pharma leadership. After earning her PhD studying hepatitis B in Vanuatu, she left academia for biotech and then Big Pharma, rising to Senior Vice President of Global Pharmaceutical Strategy at a top Swiss company. A pioneer in personalized medicine, she recently crafted a national implementation plan for precision healthcare in a G7 country. Now an author, Betsy's debut medical thriller, *The Goldilocks Genome*, spins complicated scientific concepts into a gripping thriller. Her memoir, *Mud, Microbes, and Medicine*, chronicles her unconventional path from tropical fieldwork to the biotech boardroom. A native of Berkeley, California, she resides there today.

Author photo © James Studio Photography